The Aerosol Solution

-A Novel-

by

Lee Del Mar

A'a Press

Lee Del Mar
Dr. Stuart Johnson
PO Box 2768
Aptos, CA 95001

See back of book for order forms.
A'a Press publishes fiction dealing with scientific approaches
to problems that threaten humanity and the biosphere.
a'a: Hawaiian term for basaltic lava with rough, jagged, surface.

Cover art by Frank Chlarson
Photography by Randy Johnson

ISBN: 0-9659542-1-8

Printed in the United States on recycled paper

This story is fictional.
But all parts are possible.
Some we hope will happen.
Some, never.
Sort out what you prefer
and design your future.

It is difficult for Westerners to conceive an *acceptable* method of population control that would have a chance of working.

- Garrett Hardin, *Living Within Limits*

As the *Homo sapiens* population swells, surging past the 6 billion mark at the millennium, the opportunities for pathogenic microbes multiply.

While the human race battles itself, fighting over ever more crowded turf and scarcer resources, the advantage moves to the microbes' court.

- Laurie Garrett, *The Coming Plague*

PART I

COLLEEN JAMESON: VIROLOGIST

1

LUCK

Colleen opened the door to see her friend's unhappy face dripping with rain water.

"I can't stay," the woman said. She thrust a small paper sack into Colleen's hands. "You did promise, remember?"

The university bell tower chimed 11 p.m. as Colleen unlocked the lab. She wished she had not agreed to help her friend. She often worked this late, and usually felt uneasy. The sterile white room was packed with instruments that cast distorted shadows outside the circle of light around her desk lamp.

Colleen shook a condom from the paper sack onto a lab dish and transferred a bit of the cloudy fluid to a microscope slide. Then she dropped a cover glass in place and focused the microscope. The normal symmetry of the first sperm cell reassured her.

Then a drifting cell appeared, its tail unmoving. A misshapen cap crowned the head of another. Others showed heads bent at the neck.

Colleen frowned, wiped condensation from her glasses, and rotated to a higher power lens.

Fragments of sperm cells drifted in the semen, broken pieces of nuclei, tails, and smaller unrecognizable bits. How did this happen?

Suddenly she looked up, eyes wide. She quickly withdrew rubber gloves from a dispenser and pulled them over her hands. "Sloppy," she muttered, angry at herself.

After photographing the damaged cells, she preserved the remaining semen, washed carefully, and left the lab.

On the way home, Colleen tried to think of different ways to tell her friend:

"Your husband is sterile."

2

PRIZE

Three months later, a fellow virology graduate student described a new academic prize to Colleen.

"That doesn't sound right. Are you sure?" Colleen stared at him.

"Go see for yourself," Jonah said. "*The Journal of Human Reproduction.*"

Colleen turned and ran up the stairs to the graduate library. She pressed her card into the library turnstile and found the journal. She thumbed her way to the back, to a half-page statement:

"The M. L. Birnham Foundation announces a new prize of $20,000 to be awarded once every six months. This on-going prize will go to the person who during that time describes a biological or pharmacological agent with the potential to interfere with human reproduction for three or more years."

"It's true," she gasped. "Students can apply." She scribbled an address and ran out of the library.

3

COLLEAGUES

The squeal of bicycle brakes jarred Colleen.

"Wake up. You'll get run over." A half-amused young man stopped his bicycle inches from Colleen.

"Sorry." She shook her head, and set out for the Virus Lab, her mind elsewhere again. Colleen was too harried to enjoy the bright sunlight, the crisp wind, or the flowers along Strawberry Creek.

Six weeks earlier, Colleen Jameson applied to the Birnham Pharmaceutical Foundation for the $20,000 prize. By now, many other pharmacology and virology professors and graduate students around the country hoped to win.

Even though she could barely pay for her lunch that day, she didn't want it for the money. She wanted it for prestige, weary of being invisible and ignored, the only female in the Virology Department. The director, Dr. Shevlin, avoided eye contact, and called her "Miss" Jameson.

She strode with long steps, hands in her pockets clenched into fists. A frown on her face, she ignored the students around her. Inside a coarse-knit bulky sweater that concealed her figure, her shoulders hunched from the cold. She thought of herself as drab.

Colleen also wearied of having to break into the theory discussions. Some students read little and couldn't connect their comments to the literature. Worse, some played to Dr. Shevlin's biases.

She tired of being on the defensive, of putting up a front of being a tough bitch. If she won the prize, she could forget all that.

If I don't win the prize, she worried, I'll have to kiss Shevlin's ass and maybe go to med school with all those yuppies. Shevlin might try to stop me because he knows my work's going well. I have to win it, she muttered to herself, but would they really give it to a student?

"Dammit." She overshot the room where her study cubicle competed for space among a half-dozen others. She wheeled around and headed back down the hall.

As she entered, she thought, Here we go again. She slammed the door and from the corner of her eye, saw Sam "Jonah" Jonas wince as the numbers bounced around on his digital readouts.

"Sorry, Jonah." Why do I treat him like that? she wondered. He's the only decent person in the building. "Good morning, gentlemen," she said coolly.

A student named Michael Bearing turned to her. "Colleen, could I introduce you to a psychotherapist I know? A few treatments and she might turn you into a human being. Maybe even a woman, eventually."

Tall, balding, and older than the rest, Michael had returned for a doctorate after some years in industry. He talked down to Colleen.

"I didn't know that you liked women, Bearing," Colleen said.

"Certainly I like women." He beamed at the men in the room.

They all knew Colleen studied human reproductive physiology, an irresistible target for their banter.

"I hear you can't get sperm samples for your lab work, Colleen, and I can understand why." Michael played to whoever watched, an eyebrow lifted, his forehead wrinkled into a leer. "So, in the interest of science, I've decided to offer my services." He chuckled, proud of himself. He recently gave up smoking and his idle hand clutched at his chin.

"Well thank you, Bearing. I do need some pathological material for comparison. Drop by my apartment tonight, about ten."

"What do you mean pathological?" Bearing asked. "And why ten?"

"Well, this compulsive strutting in your later years suggests genetic inadequacies, so I'll run a chromosome count." She smiled at Bearing. "I take left and right matched specimens directly from each testicle with a hypodermic needle to avoid bacterial contamination. If you come at ten, Jonah will be there to help hold you down." Colleen tossed her lab notebooks on her desk and left for the restroom. Poor Jonah, Colleen thought. He doesn't even know where my apartment is.

When she returned, she saw Michael Bearing examining one of her notebooks.

"Bearing, what are you doing?" she said. "Put that down."

He shrugged and dropped the notebook on the desk. "Don't get in an uproar. I just wondered how you're doing."

"You know lab records are confidential," she said. "If I ever find you in my notes again, I'll drag you back to the University ethics committee. After your last visit they know you don't have any."

Bearing recoiled. "Who gives a damn about your research, Jameson. Your dissertation defense is a year overdue. You'll never finish."

Without glancing up from his instruments, Jonah said, "Bearing. Shut up. You bother me."

Michael looked at Jonah, then at Colleen. "Well screw both of you." He marched out.

FACULTY

The students waited for Full Professor and Department Chair Jared Maynard Shevlin, PhD, MD to arrive at his Seminar on Advanced Topics in Virology. They crowded knee-to-knee into the stuffy, closet-sized conference room adjoining his spacious office. The department secretary often interrupted the seminar for telephone calls. Shevlin told other faculty members that this brought the students closer to his own work and broke down barriers between them. They could hear his muffled voice on the telephone in the adjoining room.

While they waited, one student worked through a stress reduction exercise. Jonah underlined a photocopy of a journal article. Michael Bearing puzzled over a new patent law for genetically engineered variants. Colleen recalled the final pages of her application to the Birnham Prize Committee.

Finally, Shevlin swept in. "Sorry, but the Johnson and Johnson Foundation called. They have an extremely large endowment." As he sat down, he lifted his trousers to preserve their knife-like crease, arranged his notebook and pen in front of him, and swung a speculative gaze around the table.

"Simmons missing again? Anyone know why?" Suddenly, no one knew Simmons. "Well." Shevlin leaned forward and made a notation in a small notebook.

He wore an informal jacket and leather sandals with expensive wool socks. He destroyed the collegiate effect, however, with a military crewcut and a severe expression. He wore Harvard and Stanford alumni rings.

"Shall we begin?" Swiveling his head around the circle of averted eyes he settled upon Colleen.

"Perhaps we can start today with Miss Jameson's sperm physiology project. You don't mind a presentation on short notice, Miss Jameson?" The other students glanced up with relieved looks.

It's Ms. Jameson, she fumed inwardly. Ordinarily, Shevlin gave students advance warning of seminar presentations.

"You feel prepared, I take it, Miss Jameson?"

"Yes, I think so." Students seldom found a topic they knew more about than Shevlin. His comments often astonished his colleagues with the depth and breadth of his reading.

"Let us hope so. Perhaps you should start with a review of the anatomical and physiological setting of your project."

Not knowing how much detail he wanted, the review threw Colleen off stride. As she described the sequence of meiosis, storage, transportation, and added metabolites, she skipped some steps on which Shevlin chose to focus. His frown deepened.

"Miss Jameson." He tapped his fingers on the desktop, and gazed out the window. "Perhaps you should make this presentation next week. I'd like you to provide more detail about this process you think you can interrupt."

That I can interrupt? she repeated to herself. But that's how the Birnham Foundation talks about the prize. Was he criticizing her or did he refer to the Birnham Prize goals? Shevlin continued to frown.

"Before we leave the topic, Miss Jameson, I'd like to ask you, as a student of physiology, whether you consider your project worthy of serious study?"

Colleen's cheeks colored. "Of course I do . . . "

Shevlin broke in. "Or do you merely seek expense money from the Birnham Foundation?"

How did he find out that she had applied? Colleen wondered. "I'm not short of expense . . ."

Again Shevlin broke in. "Perhaps you have ideological motives?" He leaned forward.

"Political?" His eyebrows arched upward as he peered at her. The students sat in brittle silence. They had never seen Shevlin so incensed.

"Perhaps your efforts represent some misshapen protuberance from the body of radical feminist thought, Miss Jameson?"

Unnerved by his intensity, Colleen faltered. "Dr. Shevlin, I haven't really thought about my motives. The Birnham Prize should interest any graduate student. For a student of pharmacology or virology, it seems like the equivalent of the Nobel Prize."

"Ah." He pounced. "So now you tell us your work is worthy of a Nobel Prize?"

Colleen gasped. She glanced toward the other students, who avoided her eyes.

"I referred to the prize, not my work."

"I see." Shevlin turned to the rest of the group. "Did anyone else consider modifying their research efforts to match the goals of the Birnham Foundation prize?"

Jonah roused himself. "Of course I thought about it. But I couldn't see any way to reorient my current work in that direction.

"Well, it seems we have two Nobel aspirants."

You word-twisting bastard, Colleen thought.

Shevlin stared at the circle of averted faces. "Who else considered applying to Birnham?" They all had speculated among themselves. Now they cowered.

Colleen recovered her courage. "We have population pressures throughout the world, and we need better methods to interrupt human reproduction."

"Oh, that's what you think?" Shevlin glanced around the group, but no one wanted to take up the argument.

"Perhaps no one saw today's Wall Street Journal." Shevlin read: "Since the Birnham Foundation Prize appeared, a dozen working scientists and a few students submitted applications." He looked around the table.

"In my opinion, researchers with demonstrated competence deserve primary access to a prize of this importance. If a student were to apply, it should be as a co-applicant under the senior faculty member who provided the greatest input into the project. Let us hope none of you applied." He scowled at them over his rimless bifocals.

"I find disturbing the expectation of monetary gain from student research."

Shevlin's expression cleared as he stood and gathered his materials. "That's enough for today." He stalked out of the room.

The students fled, except for Colleen who sat muttering to herself, "How did he find out?"

She barely noticed Jonah's hand on her shoulder. "Would you like to come over to my place? I have some cheap wine, and a loaf of bread, but no thou."

Beyond humor, she let him lead her back to pick up her lab notebooks.

4

JONAH

The higher one lives in the Berkeley hills, the higher one's socioeconomic class. Samuel "Jonah" Jonas' apartment stood only a few blocks uphill from the campus. He and Colleen looked a pair: brown hair; medium height; wiry joggers build; and worried expressions that said they had too much to do. They walked fast and energetically pushed off each step with their toes. They were preoccupied, incapable of slowing down. Typical graduate students, they hardly noticed San Francisco's glittering night lights that began to appear.

"What did the bastard want?" Colleen mused as she entered Jonah's apartment. She sat staring on the narrow bed, without a glance at his desk, radio, and computer, or the tiny refrigerator, hot plate, and handful of dishes washed in the bathroom sink. When Colleen shivered, Jonah wrapped a blanket around her and leaned her against the wall, tucking pillows on each side.

"I'll make some tea," Jonah said.

"I broke some unspoken rules today," Colleen said, "like, 'Don't argue with the professor in class.'"

"And don't overshadow your professors," Jonah added.

"Offer advisors first or second authorship of any articles you write," Colleen said. "I'm in it now. I documented my discovery with the information I put in the Birnham Foundation Prize application, and concealed it from Shevlin." She shivered. "You think Shevlin might try to throw me out of the doctoral program?"

Jonah sat beside her and handed her a steaming cup. She sipped and sniffed at the tea and seemed to revive a bit. She still sounded angry.

"That miserable hypocrite. He wants the prize money himself." She looked around the room and turned to Jonah. "Was that wise, to stand up for me today? It may cost you."

Her direct gaze and obvious gratitude warmed him. "I didn't mention it before, but Bearing often pokes around in your lab books."

"Oh god." Colleen clutched her stomach, leaned forward, and moaned. "I wrote the prize application to the Birnham Foundation from those lab notes."

"So Bearing told Shevlin about your application?" Jonah asked. A worse thought struck him. "You think he might have applied as well?"

"Maybe," Colleen said.

Jonah contemplated the possibility that a light-weight academic mercenary like Michael Bearing could apply for the prize using Colleen's data. "Those bastards," he said, wincing at Colleen's pain. Jonah went to the wash basin and soaked a hand towel with hot water. He sat close to Colleen and wiped the day's emotional grime from her face. Her eyes followed his, and by the time he finished toweling her hands, he felt her relax.

"How far are we going with this washcloth?" she asked.

"Ever had a bed bath?"

"No."

He rinsed out the towel and raised her sleeve. Rubbing with the warm, damp towel, he worked his way from her wrist toward her armpit. She took off her sweater, and his hand brushed against her breast while he washed the other arm.

"Legs are even more fun," Jonah said. He turned to the sink for more warm water. She trembled and took off her clothes under the blanket.

He felt her muscles relax as he sponged her feet and ankles. He worked his way up her legs and thighs until there was little left to wash. Colleen stopped Jonah, stood him up, took off his jeans, and pulled him down to her.

"You got me all clean. Now muss me up."

They hurried, clumsy and frantic, like their lives at the moment. They both came quickly with never a thought for safe sex, unable to slow down and extend their pleasure.

The unfamiliar warmth of a woman's body pressed against his overwhelmed Jonah. He pulled a blanket over them and clung to Colleen while they went to sleep.

They both awoke, disoriented, when the phone rang. Jonah groaned, "No one ever calls at one a.m. with good news."

A voice cracked with emotion said, "Is that you, Jonah? Thank God. I have to find Colleen Jameson. Do you know where she is? She doesn't answer her phone." Jonah barely recognized Michael Bearing's voice.

"Is that you Bearing? Why do you want Jameson?" Jonah was still inside Colleen and he began to get excited. He twisted the phone so Colleen could hear.

"Tell her she's got to come to my place. Some people here want to ask about her project."

"You want me to go to her apartment and wake her up?" Jonah slowly began to press against Colleen, making love to her again.

The answer tumbled out of Bearing. "Yes. Oh my God, yes. She must talk to these people. It may be a matter of life and death."

"All right. I'll try to find her." Jonah put the phone down, the better to enjoy Colleen.

She grabbed it from his hand. "Bearing, is that you?"

"Jameson! Thank God. You have to come over."

"I have to come over? After what you've done I wouldn't come to your funeral. Did you take any stuff, information, from my lab records?"

Silence followed. Finally, Bearing muttered, "A little. But it doesn't matter, now. Some people are here, and they want to talk to you and Jonah. Please come over. At least meet them."

"What did you do with the information you took?"

Bearing muttered unintelligibly. Jonah enjoyed Colleen, ignoring the battle.

"Damn you, Bearing, speak up. Tell me right now, or I'm going to cause you a lot of trouble. Have you ever heard of plagiarism?"

"I showed your stuff to Shevlin," Bearing said. "But it doesn't matter."

Colleen writhed with anger, and Jonah approached orgasm.

"Doesn't matter?" Colleen's voice rose. "Shevlin's ready to throw me out of the program, and it doesn't matter?"

"But that's why I called. These people are interested in your work. They appreciate it. They'll tell you, if you'll just come over."

Colleen hesitated. "Who are they?"

Jonah thought, My god, he's convinced her to go.

"They're in the export business," Bearing said, "and they need your help."

"What do they want?"

"Please, Colleen, I can't explain it myself. Their English isn't fluent, and they can communicate better in person."

"I don't believe you, Bearing, and I'm not coming."

"I am," Jonah muttered.

Bearing grew quiet. "Colleen, I can't tell you how much I regret some things I've done. But I'm in danger if you don't talk to these people." She heard his phone clatter back onto the hook.

"I hope he's in real trouble." As she hung up the phone she realized Jonah was now in the throes of orgasm inside her. She still felt warm and grateful toward him and kissed him until he quit throbbing, calmed down, and could talk again.

"Don't go near the place," he said.

"But who wants to talk to me?" Colleen asked. "I'm going with you."

"With me? I'm not going."

She dressed him and dragged him out to his car. After a short ride, she pushed him up the sidewalk to Bearing's apartment.

"There aren't any lights on. Let's go." Jonah pulled on her arm.

Colleen pressed the doorbell. "I want to hear this."

Michael peered from the door, pale, hair tousled, glasses missing. "Thank God. Come in, please." He opened the door just enough for her to enter.

Colleen faltered at Bearing's unusual disarray. She reached back and pulled Jonah after her. Jonah hoped to find Bearing drunk, but no luck.

Inside the apartment a foreigner faced them, perhaps fifty,

paunchy, with receding dark hair and thick brows. He had dark baggy eyes and a swarthy complexion. He scowled at them from his chair.

Colleen gripped Jonah's hand as they sat on the couch.

Michael said, "This is Jonah and Colleen. Uh, Mr. Smith, is it?"

The man waved Bearing aside, and ignored Colleen. "You are Samuel Jonas?" He spoke with an accent.

Jonah wondered for an instant if he should drag Colleen bodily out the door. "Yes, and your name?"

"Smith. You are a doctoral student working at the Berkeley Microbiology Lab?"

"Yes."

"Tell me your specialty."

"Mammalian reproductive physiology." Jonah described Colleen's specialty, and wondered if she would intervene. Michael looked at Jonah with a flicker of surprise, then dropped his gaze without comment. Jonah wondered, What frightens Bearing so?

"You know of the Birnham Foundation?" Smith asked.

"Of course," Jonah said.

"You applied for the prize to stop human reproduction?"

"Mr. Smith, why do you ask these questions?"

"Tell me what your application claimed."

Her territory invaded, Colleen started to interrupt.

Jonah squeezed her hand and said, "Mr. Smith, tell us your interest. Give us a reason to talk to you."

Smith's face flushed. After some inner struggle, he said, "There may be a financial benefit for you."

"Oh really? What do you have in mind?" Jonah asked.

"You found a process that will interfere with human reproduction? How does it work?" Smith demanded.

Jonah turned to Colleen, handing her the car keys. "Obviously, this is going to take some time. Why don't you take the car and go on home? I'll walk back."

"I'm staying," Colleen said.

Jonah's heart sank, and he improvised frantically. "The chemical inhibits cell division," he told Smith.

Jonah's plausible lie impressed Colleen.

"What chemical?" The man reached for a pencil.

"What would I gain if I tell you?" Jonah retorted.

"If you help us, we pay while you finish your studies. We want a description of your material and how it acts. After that, you talk to our specialists by telephone."

"How much will you pay for this?" Jonah asked.

"Five thousand dollar starting fee. Then we pay five hundred dollar every month."

"Until I finish my doctorate?"

"Yes."

"Very interesting, Mr. Smith. I'd like to think about it." Jonah started to rise.

"Tell me now if you accept," Smith said.

"Let me call you tomorrow morning. I want a little time to consider. Come, Colleen." He pulled her to her feet.

Colleen held back. "I have some questions."

"Let's go, dear." He pulled Colleen toward the door.

Smith jumped up and blocked their way. "Mr. Jonas, we must have your promise."

"It's late, Mr. Smith. Let us start fresh tomorrow morning."

"Sit down. We talk now," demanded Smith.

Jonah sighed. He wanted out. He turned and drew Colleen toward the kitchen. "Michael, we'll use your back door."

"Jonah, don't leave," Bearing pleaded.

Suddenly Smith barked, "Mahmoud!"

A taller, younger version of Smith appeared from the kitchen. He waved his hand at the couch. "You sit."

Jonah turned toward the front door and said to Smith, "Please don't try to stop us. We wish to leave."

Mahmoud shoved them onto the couch and stood over them.

"You still want to talk to these gentlemen, dear?" Jonah growled at Colleen. She shrank from Mahmoud.

"You must cooperate with us," Smith demanded.

He asked Colleen, "What is your name?"

"Colleen Jameson. It's not Jonah you should ask."

Smith's scowl deepened. "Who should I talk to?"

"Michael Bearing can tell you about the award. Ask him if he applied." Colleen's voice was grim.

Michael flinched as he heard her words. "Colleen, please don't tell them that. They'll . . ."

Smith interrupted. "Miss Jameson, we already tried to get information from Mr. Bearing. That is when he called Mr. Jonas."

Michael groaned, "Please explain to them, Colleen. We are in danger."

"Miss Jameson, tell me the chemical name."

Now, Colleen decided to follow Jonah's lead to avoid telling about her project. "It's from a chemical family that includes podophyllin, an apple root derivative. It blocks mitosis in rapidly dividing cells."

"We need a description of this material, and how you use it," Smith said.

"I could sell you a copy of my application to the Birnham Foundation," Colleen said. "It's in there."

"Colleen, don't." Jonah groaned.

"When can we get it?" Smith asked.

"I can get it from the lab tomorrow. But what will you pay for it?" Jonah hoped Colleen was trying to sound greedy. She had turned into a corporate technologist before his eyes. For a moment he wished he had left her sniveling in Shevlin's seminar room.

"This won't work," Jonah said. "The university regents take out patents and copyrights on all student and faculty developments."

Smith shrugged his shoulders. "I will give you two thousand dollars when my colleagues have looked it over."

Colleen said hotly, "I want ten thousand in cash for the first thirty pages. Then we will negotiate for the rest."

Smith leaped to his feet, shouting. Mahmoud crouched, a knife suddenly in his hand.

"Please Colleen," Michael Bearing begged. He lurched toward her, but fell against Mahmoud.

Thinking himself attacked, Mahmoud's knife flicked upward into Bearing's belly. He screamed and fell, crashing through a glass table.

Too late, Smith shouted to Mahmoud, "No!"

Suddenly a new, angry, muffled voice came from the apartment next door. "That's it. I warned you. I'm calling the police. I have to go to work in a few hours." The voice broke off with a string of curses.

Jonah pulled Colleen away from Mahmoud and opened the front door with his handkerchief in his hand. They ran toward the car leaving Smith shouting at Mahmoud.

"We have to call an ambulance," Colleen said as they jumped in their car.

"Right." Jonah tried to remember the nearest pay phone.

When a police car sped past them with lights flashing, Colleen watched through the rear window.

"They stopped at Bearing's."

Jonah drove by his apartment, parked some distance away, and stared at the door.

"What's the matter?" she asked.

"I wonder if it's safe to go in?"

Once inside the apartment they placed furniture against the door.

"How did you know that we shouldn't have gone?" Colleen asked.

"Bearing usually blusters when attacked. His panic meant he feared for his life. It scared me," Jonah said.

"Well, you warned me." She shook her head. "Can I stay here tonight?"

"Of course."

She acted as if she had been mauled and curled next to him with her back to his belly. Jonah felt his groin swell as he imagined he was inside her again. While they drifted off, Colleen spoke.

"I've never been pushed around like that. Thank you for taking care of me."

5

CHRISTIANS ONE, COLLEEN ONE

Colleen awoke slowly to find herself completely entangled with Jonah, arms, legs, clothes, hair, pillow, and blankets. She could feel him aroused as he woke up. He rubbed her thighs, nuzzled her neck, and sniffed at her hair and skin. Pheromones, she thought, enjoying the feelings.

"Are we having an affair?" she asked.

"Not yet." He rolled her on top of him, and she buried her face in his neck.

"Oh no," she wailed. The images from Bearing's apartment suddenly flooded her mind. "What'll we do?"

Jonah nuzzled her. "Even after last night, all I can think of is sex. No wonder people have so many babies."

"I'm not ready to do anything right now." Colleen avoided him as he tried to enter her. "Can you wait?"

"I'll try," he said, his face buried in her hair.

"I'm distracted," she said.

"Me too," Jonah said.

"First I'm going home. I want to change clothes, and I really want to pick up my copy of the Birnham application." She stopped. "Is it all right if I go now and come back soon?"

"Yes." He tried not to move. "Come soon."

She raised her head and looked at him. "What are you doing? Oh, you poor thing."

Although she wouldn't let him inside her, he couldn't stop. She kissed him while he had a long slow orgasm. As he calmed down, she asked:

"Are we having an affair yet?"

"I am. But you'll have to wait," he said.

"That's not fair."

"Well, if you'll wait a minute . . ."

"Never mind," she interrupted. "I'll come back."

Colleen jogged home and thought about Jonah. He differed from other men who pressed her for sex. He seemed guileless, with no approach, no line, and only talked about what he felt at the moment. She had warm feelings and resisted an impulse to run back to his apartment.

When she arrived at her apartment door, she found the landlady waiting for her. "Oh god, am I late on my rent again?" Colleen tried to remember when she paid last.

"Why did those gentlemen want to go in your room?" the landlady demanded.

"Some gentlemen wanted into my room? They knew me?"

"They knew you all right. They asked if you were the doctoral student. Are you a doctoral student?" She squinted at Colleen.

"Of course, I am," Colleen answered.

"So they knew you were a doctoral student. What's been going on in that room? I won't have anything illegal or immoral in this house."

Colleen cut her off. "You said they wanted to go in my room. Did you let them in?"

"Why shouldn't I let them in? They only wanted to inspect your room and make sure you study in the doctoral program."

"You let them in?" Colleen sputtered.

"Why shouldn't I?" Her landlady raised her voice. "They have the right. They told me they're with the government."

Suddenly, the landlady's critical expression evaporated. She leaned toward Colleen with her hand raised to hide a whisper. "Actually, I didn't let them in. But they said they would come back." She smiled innocently at Colleen.

"Oh. Oh, thank you." Colleen kissed the older woman on the cheek, and ran into her apartment. She picked up her mail and the

Birnham application. Her cat rubbed against Colleen's legs, marking Colleen off as the cat's private property. The cat also demanded attention, complaining that Colleen left her alone too much. Colleen fed her and explained that graduate students don't have much time.

As Colleen started her shower, she thought of Jonah sniffing at her like an animal, unable to resist her. Maybe I should quit showering, she mused. She once read that whenever Napoleon returned from a military campaign he wrote ahead to Josephine, "Stop bathing. I'm coming home."

When Colleen finished dressing, someone rang the doorbell. She froze, fearful that the landlady's "gentlemen" had returned. Then she panicked with another thought. Could it be "Mr. Smith." She peeked out a window and saw a young woman with a clipboard, dressed casually, student style. Colleen opened the door.

"Yes?"

"I'm a reporter for the campus newspaper," the young woman said. "I'd like an appointment to talk with Colleen Jameson. Is that you?"

"You're with the *Daily California?*" Colleen asked. "Why do you want to talk to me?"

"One of the other reporters said you're being considered for an award for your work in virology."

Colleen flushed with pleasure. "Well, I've applied for a prize."

The reporter said, "I'm Susan Brownlee, and my friend Carl is waiting outside. May we sit down and take some notes?"

Colleen looked at the man. He wore dressy black slacks, a short-sleeved white business shirt with a dark tie, and polished black shoes. He cut his hair short.

Colleen looked uncertain and Brownlee said, "Listen, I don't want to interrupt you. Shall we leave and come back another time?"

"No, come in," Colleen said. She felt less fearful with someone else in the apartment. Besides, no one had shown interest in her research work for a long time.

While Susan Brownlee drew background details from Colleen, Carl flipped open a lap-top computer and typed rapidly as Colleen talked. Pretty high-tech, Colleen thought. I wish I had a gun-bearer like that to take notes for me.

Colleen talked her way forward in time to the current project. She described only generally her work leading to the Birnham Foundation application.

Brownlee asked, "How did you get started with your study of human reproduction?"

"I became interested when I took a job studying selective breeding practices with large mammals."

"You're developing selective breeding programs for people? Is eugenics the right word?"

Colleen laughed and said, "We probably need some, but that's not why I became involved with the project."

Brownlee said, "You've undoubtedly benefited from the advances made by the women's movement. Are you a member of NOW? How about other pro-abortion groups?"

"Well, I'm a feminist. I believe in comparable pay. I want more women to go to college and break out of the get-married-and-have-babies trap. But I'm not a part of any organizations."

"If you can interrupt human reproduction, would Planned Parenthood be interested in contraceptive applications from your research?"

Colleen felt reluctant to describe her project in detail to the general public. "It wouldn't be useful to Planned Parenthood."

The reporter pressed for other applications. "How about abortion? Couldn't your work contribute to the development of new abortion procedures?"

Colleen sighed. "No, it isn't aimed at that part of the reproductive cycle."

"So it interferes with another part of the cycle. Ovulation? Fertilization?"

"No."

"Does it work like contraceptive pills or abortion pills?"

"No." Colleen became more wary of the questions.

"How do you want me to describe your research?" Brownlee asked. "Can you be more specific?"

Colleen searched for a simple description that would conceal her specific findings. "There is a chemical involved that prevents sex cells from forming." Half true, she thought.

"Ah," Brownlee said. "The chemical sterilizes? Does it work on both men and women?"

Colleen felt increasingly uneasy. "No. I don't want to discuss how the material could be used. Whatever you print could jeopardize my application to the Birnham Foundation."

"Just two more questions. Are you religious?"

"No. Does that matter?" Colleen asked.

"To some readers. And have you yourself ever had an abortion?"

"No," Colleen said, offended. "And I won't answer any more questions."

"We'll decide when to stop," Brownlee said.

Colleen heard the the computer snap shut. She looked up to find Carl staring at her, his face tense.

"Let us explain some things," he said. "Your answers to our questions show that you developed something for eugenics and abortions. You told us you support feminist anti-family groups like Planned Parenthood."

Colleen said, "You'd better leave now." She lifted the cat from her lap and stood to escort them to the door. "You aren't *Daily California* reporters. You're not students, either."

"That's right," Brownlee said. "But your research will be reported fully in our newsletters. We're part of a large network of people who believe in the Christian family. Obviously, you don't."

"Yes," Carl said. "And here is the message we bring to you." He leaned forward, and spoke slowly. "Some of you women think you are superior after you go to college for a while." He smiled.

"You must stop this Godless research and withdraw your application for the prize."

"You can't be serious," Colleen said. "I live for my research."

"You could die for your research," he said. "Your studies could jeopardize your life."

"What do you mean?" Colleen's voice rose.

"If you continue to work on ways to kill babies, you will be in danger," the woman said.

"I'm not killing babies with my research," Colleen exclaimed. "What are you saying?"

"You could lose your academic appointment or your job as a research assistant," Brownlee said. "You might endanger the well-being of your loved ones, did you think of that? Don't your parents live in Piedmont? On Clairmont Avenue?"

Colleen gasped. "Both of you leave immediately. You're trying to threaten me."

The man picked up the theme. "What happened to those professional baby killers could happen to you. You remember what happened to those doctors who worked in abortion clinics?"

Colleen's face contorted with anger. She tried to push past the man. He reached out and closed his hand around her cat's neck, and stripped it from her arms. It hung suspended from his grip, its tail and legs flailing.

"Please don't hurt her," Colleen pleaded. Blocked from the front door, she backed into the kitchen as he advanced. The cat still dangled from his hand.

“Maybe you worship Satan? Is this your familiar?” He thrust the cat in her face.

“No. Please don’t hurt her.” Colleen tried to loosen the cat from his grip, scratching him as hard as she could. Choking, the cat tried to claw free, and slashed his hand.

He cursed and slammed Colleen against the kitchen counter, holding her immobile. He increased his pressure on the cat’s neck. It writhed for a moment, then its neck snapped.

Colleen begged, “No!” She opened a drawer and found a kitchen knife.

The man jumped back and threw the cat’s body at her. “Take it.” Then he hit Colleen in the face, knocking the knife from her hand.

Colleen ran out the back door. She rushed to the end of the alley lined with high board fences, and darted into the back yard of an acquaintance. She heard them following. She ran onto the back porch, and twisted the locked door handle. Desperate, she kicked aside a potted plant near the door. A key lay exposed and she scooped it up.

The man burst through the backyard gate, a few steps away. Colleen pushed the key into the lock, jumped inside, and slammed the door behind her. She locked the barrel bolt.

While she panted and rested, she felt two hands suddenly grip her ankle. “Got her,” the woman said, outside the door. Brownlee’s arms reached through a lower “doggie” panel, holding Colleen. Carl peered through the curtain that covered the upper window pane.

Colleen kicked her legs, trying to free them.

Brownlee grunted, “I can’t hold her.”

Carl cursed and punched his fist through the window pane. Glass flew as his arm groped for the door handle. He turned it, but the barrel bolt still held. Then his hand, covered with blood from glass cuts and cat’s claws, searched for the barrel bolt.

Colleen’s anger flared when she saw the cat’s scratch marks on his arm. Then she had a vision of herself trying to fight off both of them when they broke through the door. With that, she clamped both hands around Carl’s wrist. She dropped to the floor, hung on his wrist with her full weight, and sawed his arm sideways.

Carl tried to suppress a scream as Colleen pulled his forearm down onto the jagged shards of glass that stuck up from the edge of the window frame. Blood spurted from bone-deep slashes. Arterial, Colleen’s mind noted clinically as she saw the bright red spurts. Soon he’ll be too weak to chase me.

Colleen ran for the front door leaving them to exclaim over his wounds. She jumped out into the narrow street and jogged a circuitous route toward Jonah's apartment. She paused and watched outside his front door for a few minutes. When she knocked, Jonah didn't answer. Colleen let herself in and found the apartment a shambles. She locked the door, found a hammer, pulled a chair against the front door, and sat down. She clutched the hammer in her hand and rested her head on a pillow to wait for Jonah. She wept as she thought of her cat, casually killed, dangling helplessly,

When Colleen heard the apartment door rattle, she raised the hammer ready to hit whoever came in.

"Whoa." Jonah jumped back.

Colleen dropped the hammer and hugged him. She wailed about the Christian terrorist reporters who killed her cat. Jonah made her tell him twice about dragging the man's wrist across the edges of the broken pane of glass.

Then Jonah told her that he visited the university lab. "Bearing died last night. The police questioned people all morning. They assume a drug addict killed him."

"That's all they know?" Colleen asked.

"I guess so. They couldn't have been more wrong about poor Bearing. He thought his first million dollars would come from genetic engineering."

6

DISCOVERY

The next morning Colleen awoke first. "You got ahead yesterday," she complained to Jonah's ear. "You knew I had to go, and you came anyway."

He roused himself. "Yes, thank you. I enjoyed that."

"Just like a man. Use a woman for your selfish pleasures and then leave her to her own devices."

"You use devices? What kind?" he asked.

She giggled and shook her head. "No."

"I really enjoyed outercourse yesterday," continued Jonah. "But intercourse might be more fun."

"Fun? *Might* be more fun?" Colleen said, incensed. "I'll show you fun."

AND THIS LITTLE PIGGIE BECAME STERILE

As they recovered, Jonah asked, "By now, do you think you know me well enough to tell me about your research project?"

When Colleen hesitated, Jonah pressed on.

"If not, maybe there are some other things we could do to become even more intimate? Something kinky, perhaps? Does anything special come to mind?"

"I'll think about it and let you know." She laughed. "But you're right. You can't help me with the problems we've had unless you know more about the Birnham application."

Colleen had trouble talking about her projects. Researchers guarded new knowledge jealously. Students sometimes "borrowed" ideas from each other. A student and faculty member might co-author "mutual findings." Professors picked others' brains. Faculty members obtained promotions and received research grants on the excellence of their studies, while students won appointments and fellowships.

"Did you find a chemical agent, or a microbe?" Jonah asked.

Colleen sighed. What's the matter with me? she wondered. I let Jonah have access to my body. Why can't I turn over some information? I have to trust him. Finally, she told him, "A virus."

"You want to talk about this another time?" Jonah asked.

Colleen looked at him. He seemed worried, as if he didn't want to drive her away and jeopardize their relationship.

"Let's go to the Lab." She knew she felt secure at work there. "We can talk while I process some samples."

In the preparation room outside the Lab, Colleen helped Jonah into isolation garments to cover his street clothing. "You're not supposed to be here, you know. Hold your hand over the filter intake to make sure the respirator fits tight. Can you breathe?"

He shook his head, negative.

"Good. That means it doesn't leak. Now take your hand away and make sure the respirator stays tight all around the edges."

Once inside her cubicle, Colleen inserted her arms into a plastic glove box. Now in her element, she skillfully manipulated equipment, macerating and centrifuging cell cultures.

"Where did you find the virus?" Jonah asked. The respirator and background noise from exhaust fans muffled Jonah's voice.

"From Fred, my girlfriend's husband. She asked me to check his sperm because she couldn't get pregnant."

"You found something wrong?" Jonah asked.

"Everything. Inactive sperm, deformed sperm, sperm fragments. I told her to trade him in." She laughed.

"No you didn't."

"You're right. But I might as well have. They split up and we can't find him now," Colleen said. "Anyway, I froze and saved the rest of his semen."

"What made you go back to it?" Jonah asked.

"I kept seeing those pieces of cells. They looked like exploded fragments. And that made me think of lysed cells, burst as the virus multiplied inside them."

"So, how did you show it carried a virus?" Jonah asked.

"I had access to a swine selective-breeding project. I isolated two non-breeding boars, filtered the semen carrying Fred's damaged sperm, and smeared the filtrate onto the penis of one of the boars."

"In case it might be transmitted sexually?" Jonah asked.

"Right. Then I diluted the filtrate with water and sprayed it into the nostrils of the second boar."

"On the assumption it might be an airborne virus," Jonah said.

"Right. Semen samples showed infections in both boars; no sperm cells present. After that, I just followed the Koch postulates and filtered the inactive semen of the infected boars. This time I applied the filtrate as an aerosol to two new boars. Again, the filtrate sterilized both boars and yielded semen without active sperm. I had one boar slaughtered, and took tissue specimens from the testes. Scars sealed the seminiferous tubules. I have photographs of those tissues and the slides."

"So the virus blocks the semen delivery tube? It's a viral vasectomy," Jonah said.

"Not quite. The sperm production tubes become sealed, not the delivery tubes." Colleen said.

"The boars stayed infertile?"

"I took repeated samples from two of them, and never found live sperm. I'm pretty sure it's permanent. The tubes I dissected were badly damaged."

"You propagated the virus through pigs. Will it still infect humans?" Jonah asked.

Colleen eyed him speculatively. "I don't know. I haven't had a male volunteer yet."

Jonah stepped back a pace.

"You're safe. I'm not through with you, yet," Colleen said. "Anyway, I'd sure hate to bet it won't infect humans."

She unloaded six tubes from the centrifuge and decanted a clear fluid from each. She capped the six new containers and wrapped them in plastic sheeting. She held up the samples of fluid. "Whatever else happens, I have the virus."

"And you have a paper trail to prove priority of your findings. Now how do we keep the virus strain in your hands only?" Jonah asked. "When I say 'we,' I assume you want me to help you?"

"Yes, I do want you to help," Colleen said. "I've been thinking about that. Let's separate these vials and freeze them, maybe in fish carcasses, or put them in frozen food cartons to hide them. I know they'll last several months that way. We could also put them in a public freezer locker and pay a year ahead."

"Will they stay infectious if they stay frozen?"

"They have so far. We could also ask some friends to keep a package in their freezers for a while."

Leaving the lab, they peeled off the protective clothing and carried out the virus samples, glancing behind them to locate any followers.

Later that night, as they went to sleep, Jonah asked, "Do you think any of that stuff got on me?"

"What stuff?"

"Your virus."

"Could be," she threatened.

The next day, Jonah made up samples of his sperm and preserved them along with the sample bottles of virus.

7

REGIS JAMESON, ESQ.

Jonah and Colleen felt vulnerable. They worried about someone breaking in while they slept. They also wondered how the foreigners got Colleen's name.

"Why don't I go find another apartment?" Jonah asked.

"How about a boat?" Colleen countered.

"A boat? You have a boat?" Jonah demanded.

"My brother keeps a sailboat down at the Berkeley Marina," Colleen announced.

"How big is it? How many bunks?"

"Is that all you care about?" Colleen asked. "Whether it has a bed?"

"No," he said. "It's a quick way for amateurs to tell how big it is."

"Four or six. I'm not sure. Why do you say I'm an amateur?"

"That's big enough to live on," Jonah said. "Would he let us use it?"

"Maybe, if I ask him right. Let's call him."

She gave Jonah a telephone number to dial. He heard the receptionist say, "Rosehill, Whitfield, and Jameson." When he handed the phone to Colleen, she asked for Regis.

"Your brother's a lawyer?"

"He's an attorney," Colleen said.

"Close enough."

When Regis Jameson came to the telephone, his sister explained that she and her friend wanted to move aboard Regis' boat and that she had some problems.

"Sure, you can use the boat. Who is your friend? What kind of problems?"

"Thanks, Regis. We'll keep it in the slip. My friend is a very nice man, a fellow virology student. I think I'm living with him if we can find a place to live."

"So what about the problems? Something he did to you?"

"He did lots of things to me, all nice. I'll have to tell you the problems in person. What's your hourly fee sir?" At the last moment, her tone shifted to a dependent worried sister.

"If things get complicated, my fee could be very high," Regis said. "I'll come by the boat first thing in the morning."

Jonah was an old hand with boats but hadn't been on one since he entered the doctoral program. He and Colleen drove through Berkeley's commercial section toward the East Bay freeway and saw the unmistakable forest of sailboat masts. Jonah shivered with anticipation.

Once aboard, the size and accommodations of the boat dazzled him. "This boat can go to Tahiti," he told Colleen. "It's bigger than some apartments I've had in Berkeley."

Colleen lay down on the only double bunk. "Not much room," she said. Jonah clambered in after her.

"Notice who is thinking about the bed, now? You're right. I almost have to lie on top of you."

They moved aboard that day. As they settled in, Jonah called her to the steering station. "Look what I found in the map locker."

"Hey, my brother's .22 rifle." Colleen took it, checked the empty chamber and full magazine, then pressed the safety to 'off.' She handled it expertly.

"You look like you know how to use that," Jonah said.

"Regis taught me," Colleen said. She pointed to a small sharpshooter's trophy amid several larger ones. "That's mine. My brother won the others."

"Really? He's a competition marksman?"

"Won his first match at twelve," Colleen said. "He's also chairman of the Western Chapter of the NAR."

"You mean National Association of Riflemen, as in 'everyone should own a gun and go hunting?'"

"That's right," Colleen said.

"I don't think I ever met a diehard Rifleman."

"Well, you'll meet one tomorrow," Colleen said.

That night they felt more secure aboard the boat.

"Every once in a while I just want to go to the police and ask for help," Colleen said.

"You want to explain to the police that Michael's killers really were after you?" Jonah asked. "And why they wanted you?"

"No. I don't even like to talk to the police about a parking ticket," Colleen said.

"At some point they'd find out about the Birnham Foundation application," Jonah said. "They'd bring in the Feds, and confiscate your research samples, and you'll never get to finish documenting the virus or its effects. You could be knocked out of biological research and not get your degree. Maybe it's too late already."

"But if the virus spreads and becomes a familiar sexually transmitted disease, I'll feel we should have warned the authorities."

Jonah noticed that she now included him in her project. "I don't have to remind you that you found the virus in a member of the general population when your girl friend brought you her husband's semen sample. His name was Fred?"

"Yeah, Fred," she said.

Colleen looked so depressed Jonah changed the subject.

"Can you think of anywhere to run if they come after us on the boat?"

"As a matter of fact, I can. My folks have a trailer up north in the redwoods. I talked to them and told them some of what has happened. I told them we'd be on Regis' boat or up in their trailer, and not to tell anyone. They sounded pleased."

"That you have murderers chasing you?" Jonah asked.

"No, that worried them. But when I told them I'd moved in with you, my mother said, 'Good. I was worried about you.'"

"Just like that?" Jonah asked. "No security check for me?"

"I told them all about you," she said.

That night she coaxed him off to bed early.

"Have you always been like this?" he asked, after a long, sensual evening.

"Like what?" she asked.

"So hungry."

"Sexually hungry?" She thought about it. "I don't think so. I haven't slept with many men."

"Do you want to take some precautions so you don't get pregnant?"

"No! I won't!" Her outburst startled her.

He stared at her, then pulled her to him. "I think you're marvelous and I want to live with you." He kissed her hard for several moments. "Do you want to have a baby?"

She buried her face in his shoulder. First she said, "Maybe so." Then she shook herself and said "No!"

He kissed her and rocked her for a while.

Finally, she said, "I'll see how I feel tomorrow."

The next day she telephoned a friend, a graduate student studying endocrinology. They discussed the effects of the available contraceptive pills over time. Then Colleen went to the university clinic and had an intrauterine ring implanted. She was mechanist to the core.

During the examination, the gynecologist asked, "You're off your menstrual cycle. Do you want to be regular again?"

"Yes," Colleen said absently. Thirty minutes later, on the way home, she realized she had just had a vacuum-aspirated abortion.

8

SAFE

The sea breeze chilled and dampened the bedclothes in their bunk. Jonah awoke with feelings from his childhood on the ocean. He got up, turned on the galley stove to warm the cabin, and made coffee. Watching Colleen wake up, he realized how much he liked being with her. He sat on the edge of the bunk with her cheek in his hand.

"I should tell you about the good feelings running around inside me, watching you sleep."

She turned her head and kissed his palm. "I'm glad." She lifted the covers and pulled him into bed. They held each other for a long time, feeling the boat rock, listening to the waves slap the hull, and the wind bang the ropes on the mast.

"I feel safe when you're with me," he said.

Colleen raised her head to look at Jonah. "But you're not."

"Not what?"

"Safe." She rolled on top of him, reached down and placed just the tip of his penis inside her. Then she held him down and teased him with her tongue. He sighed. She kissed and bit him for several minutes. Jonah moaned, and tried to complain. She stopped him from talking with her mouth on his. She gently moved her hips. He begged. She ignored him as he whimpered, his breath heaving continuously. She indulged herself, murmured over him at her leisure, pressing him into her a bit, then pulling back. She trembled occasionally as her excitement rose and waited a long time before she let the waves of heat that washed over her sweep them away.

Colleen stretched and breathed deeply as she lay on Jonah's heaving chest. She arched and flexed her tingling body. Jonah slid out from under her and went to the washroom. When he returned, he immediately started breakfast.

At first, they ate wordlessly.

Finally Jonah announced, "I could have had a heart attack, you know."

"Poor dear." She leaned over and gave him a toast and marmalade kiss. Then she licked a crumb off his lip. "Quit your complaining."

When Jonah regained his composure, all thoughts turned back to the boat.

"I feel like taking a south seas cruise," Jonah said. "We'd be safe down there for a while."

"We might be safe, but we'd be out of jobs in microbiology," she said.

"There's always work for technical types like us," Jonah said.

"Sure, examining stool specimens for parasites, and running medical labs," she complained. "Besides, we'd never get back to our doctoral program."

"Well, we'd have each other."

Colleen stopped and stared at him. "That's supposed to be the woman's line, Jonah. I want my doctoral degree, with honors, and with the research job offers that go with it!"

"Well, do you really think you're going to get it from Shevlin? He's going to try and drive you out of the program. He's already said that your dissertation project isn't scientific enough."

"The Birnham Prize could turn that around," she muttered.

"So we go back and dodge thugs at school, trying to finish our doctoral programs?" Jonah shouted. "You're not serious."

She quieted him with kisses. "Can't shout in a boat harbor. The Coast Guard might come."

Jonah took Colleen on a tour of the boat. He showed her how to shut down the electrical system, where to find the running light switches, how to start and shut down the diesel, how to flush the toilet,where to attach the pump out hose, how to start the generator, where to bleed the fuel lines, and where to fill the tanks. He pointed out that they needed filling now. Colleen liked all this, since Regis never let her run the boat.

Regis arrived as they finished the tour. He was taller and bigger boned than Colleen, short haired, fair, and a bit burly. He looked uncomfortable in his suit and tie.

Colleen kept him talking with Jonah about Regis' boat, Jonah's boats, boat motors, sail rigs, anchors, and all the trips they would take when they were rich and famous. When she decided the two men approved of each other, she told Regis about the attacks on them.

Regis asked about Michael Bearing and Shevlin's hostility.

"Could they be working on their own Birnham prize application?"

"Possibly," Colleen said.

"Bearing often criticized Colleen and her work." Jonah said.

They told him about the visit to Bearing's apartment, the threats of the foreigners, and Bearing's death. Colleen described defending herself from the religious terrorists.

"This is very serious." Regis said. "Why didn't you call me?"

"We've been running most of the time," Colleen said.

Regis asked questions to find out what the police knew of Colleen and Jonah's involvement.

"You'd better let me get you a lawyer to help you notify the police."

"I don't want to talk to the police," Colleen said.

"I'll bet you don't," Regis said. "How about these foreigners? Why did they come to Michael Bearing?" Colleen and Jonah shook their heads.

"No idea."

Regis thought for a while, then told them, "I'm going to the office. Would you stay here until I can call you in an hour or so?"

As he hugged Colleen goodbye, she tried to thank him and burst into tears. "The bastards are getting to you, huh? Never mind. Let me make some calls." He shook Jonah's hand. "I'll get back to you as soon as I can."

Two hours later Regis called, full of news. "I talked to the director of the Birnham Foundation." Colleen panicked, fearful his call would jeopardize her application. Jonah and Colleen pressed their ears to the telephone.

"He likes to talk," Regis said, "and he remembered your application. The Wall Street Journal article generated a lot of outside interest. He sounds proud of the Foundation's decision to offer the Prize."

"Now comes the surprise," Regis said. "He told me an Iraqi businessman in San Francisco called and asked for the names of some of the researchers. The Director gave him a few local names including yours. I think he told everyone who inquired. The Journal article was very good advertising for them."

"You mean he selected my name to give people who called them?" Colleen couldn't believe her ears.

"That's what he said. And there's more," Regis said. "Ever hear of Biosolz Corporation?"

"Big producers of pharmaceuticals?" Jonah asked.

"That's the one," Regis said. "They called the Birnham Foundation and asked for some names of applicants. The Director told them he liked what you did and gave them your name."

"He liked my application?" Colleen asked. She had been punished for months by Bearing, then Shevlin. While she tried to recover, Regis told them:

"Now I want to call Biosolz Corporation. Talk to you later."

"Why?" Colleen tried to ask. "He hung up."

"So that's what lawyers are for. I mean attorneys," Jonah said. "What a helpful guy."

"It's a good thing he didn't tell me what he planned. I'd have forbidden him to call Birnham. Not that he'd have listened," she added.

Curious and agitated, they talked for an hour until Regis called.

"Quick, what?" Colleen demanded.

"I do believe Biosolz wants to offer you a job," Regis said.

"Are you serious?" Colleen shouted.

"Yes. I'm having dinner with two local company representatives tonight. Are you interested? And do you trust me to gather more information, before you talk to them?"

"I am. I do. I love you. Please act as my attorney and representative," Colleen said. "Oh, by the way, tell them that I only come as a team, with Jonah."

Regis laughed. "Ah, new love. OK, whatever you say."

Jonah got on the phone and asked Regis if he would like them to fill the boat's fuel tanks. Regis said "Sure," and told them how to get to the fuel dock.

Colleen paid close attention when Jonah had her start the engine and steer the boat away from their position at the end of Regis's dock. She counted as she passed the eight piers on the way to the fuel dock. The crowd of waiting boats forced Colleen to steer into a bank of empty slips nearby. Jonah stepped easily onto the pier, and snubbed a line on the nearest cleat. He drew his finger across his neck to tell Colleen to cut the motor. Jonah fastened the forward mooring line and rejoined Colleen at the boat's helm.

"So what's so hard about skippering?" Jonah asked. "You handled the boat well." Colleen took a deep breath and stood straighter.

While they waited for the fuel dock to clear, Jonah went out for a pizza and a bottle of California champagne. "Nothing but the best for my girl," Jonah said. "How do you open these anyway?" He pried at the wire on the bottle.

"How would I know?" she said. "I'm just a grad student."

They ate, drank, and speculated about various futures.

"Where is Biosolz?" Colleen asked.

"North Carolina, in the Research Quadrangle."

"That's a high class place," Colleen said.

"You've been there?"

"Drove by once," Colleen said. "Hey, look. They closed the fuel dock. Can we keep the boat here tonight and fuel up in the morning?"

"I think so," Jonah said. "The place is deserted. No lights in any boats, no live-aboards, no barbecues or cocktail parties. No executives revving their engines to impress their secretaries."

"C'mon then. I'm sleepy." Colleen seldom drank champagne.

"Wait," Jonah said. He balanced an empty beer can at the top of the stairwell into the galley.

"What's that for?" she asked.

"Burglar alarm," Jonah giggled.

9

THE OK MARINA

The downward tilt of the port side of the boat gave the first warning. The hatch slid open, and the beer can bounced to the galley floor.

Jonah scrambled out of the bunk, naked, into a flashlight beam close to his face. A man dressed in black with a stocking cap pulled over his face held the light.

"No move. No noise," he whispered, with the now familiar accent of the men in Bearing's apartment. Another black-clad figure dropped through the hatch, and stood over them. "Give me guns."

Jonah said, "We don't have any."

Colleen sat up and the second man immediately hit her hard with the back of his hand. The blow knocked Colleen's head against the bulkhead, and gashed her scalp. She sprawled flat on the bunk.

"Guns quick." The dark figure grabbed Colleen by the hair, blood streaming down her face.

"No noise. I hit her again. Give guns."

"Please," Jonah said. "Don't hit her again." He pulled the blankets and pillows away from the bunk with two fingers. "Look. No guns."

The man dragged Colleen into the galley and threw her onto a berth where she lay unclothed and unconscious. "You tell us quick. If no, we do things to her." The man removed his cap. He smiled and put his hand between Colleen's thighs, caressing them roughly. She moaned slightly.

"Ha! Look. You see? She likes." He ignored the blood on her head and shoulders.

The man stood six feet, with a pockmarked face, jet-black hair, and a piercing stare. Jonah cringed as the man leaned close to Jonah, and hissed through rotted teeth. "This stuff you make so people have no babies?" He cut imaginary tissue with a scissors motion of his fingers.

"You give us, now." To emphasize his request, he went to Colleen's dangling head, seized her by the hair, and drew back his fist to hit her.

"I'll give it to you now," Jonah opened the freezer compartment of the refrigerator and dug in the ice chest. He unwrapped a package and showed them the bottle.

The large hooded man discarded the wrapping, shook the bottle, held a flashlight behind it, shrugged, and put it in a leather purse around his waist. The two then pulled Jonah up on deck, leaving Colleen collapsed on the bunk.

As they stepped off the boat, the overhead dock floodlights revealed the man Jonah saw in Bearing's apartment.

"Mahmoud," Jonah gasped.

"Yes, bastard." The large man whipped off his stocking cap, and stuck his face close to Jonah's. "You remember good." Mahmoud turned to the other three. "Kill him. Then come." Mahmoud raced off toward the shore with the bottle in his waist pouch.

Two of the men held Jonah's arms while the third lifted a knife toward his neck.

A ragged voice suddenly shouted from behind them, "No more!" Three quick shots hit the knife wielder in the head. He fell backward into the water.

Jonah saw Colleen on deck, holding Regis' rifle, a guttural shout contorting her face.

"Allah, Ach'bar. I am woman." Naked, in a towering rage, she bellowed her enemy's own battle cry. Blood smeared her white face and torso, and dried in her hair to form spikes, a gory punker vision.

She fired another volley and the man next to Jonah staggered away. The third man cringed behind Jonah. Screaming the same battle cry, Colleen shot him first in the leg, and when he fell to the dock, she dispatched him where he lay. She emptied the rifle at Mahmoud but he reached the van and fled.

Jonah rolled the man at his feet into the water, wondering, How did she shoot him while he was hiding behind me?

Colleen stepped to the boat's steering wheel and waved the rifle at Jonah.

"Come, Jonah," she said calmly.

Jonah freed the dock lines and Colleen started the diesel. He scrambled aboard as she cranked the wheel, pulled away from the dock, and headed like a moth for the brightly lit Bay Bridge.

Looking at Colleen, Jonah thought, It must be strong medicine to kill your oppressors. Her warrior woman's body bare and upright, Colleen smiled, staring at the far lights, rifle in one hand, boat's wheel in the other.

Jonah approached Colleen. Who is this woman? he wondered. "How did you get the rifle in action so fast?"

She turned, her eyes glowing. "They tried to kill you. And that son-of-a-bitch hit me. I'll bet he won't do that again."

"Colleen, he's probably dead."

"Good for him and the other guy too. He stuck his hand between my legs."

"Colleen." Jonah shook her shoulder. "Tonight didn't happen. They can trace the gun to Regis."

"Ah." After a pause, she said, "All right." She looked at her brother's childhood companion, his trusty .22 rifle which had just killed three people and saved Jonah's life, and threw it into the harbor water.

With the boat back in its slip, Jonah applied a butterfly patch to Colleen's scalp cut. They went to bed to await the police.

But the police never came.

10

HIGH DESERT

When Colleen had first arrived as a new student at the university, she felt claustrophobic in crowds. From age twelve through seventeen, she attended a live-in private school for children whose parents could afford the fees.

This prestigious school with its own dude ranch occupied 200 acres on the Colorado plateau near the Rocky Mountain foothills. Teachers on horseback supervised the students as they explored the high desert and studied the wild life of the region. Colleen's father made sure her teachers emphasized the physical sciences and mathematics.

Not only did she learn to ride horseback, she shot skeet, fished, and even hunted. After she shot a deer, Colleen unnerved the cowboy guides when she made them show her how to dress out the carcass with a buck knife. She made sure the body parts looked like the diagrams in the books she studied. Like a good student, she turned the event into a clinical dissection.

When puberty struck, most of the students traded their horses for sport cars until they graduated. Colleen raised a few eyebrows when she asked for a pickup truck and spent her weekends camping in the desert. She developed a kinship for animals and wilderness areas.

Each summer, the students selected a project, a real-world investigation related to their classroom studies. In her last year, Colleen arranged with the Bureau of Indian Affairs to spend time at a uranium mining site on the Navajo Reservation.

She sampled the residual ore piles, and traced the path of pollution carried by water percolating through them. She searched out the remaining miners, some with lung cancer. She talked with them at length, and what she learned disturbed her.

Colleen had long conversations with her father and demanded explanations as to how the mining companies could show such carelessness for the health of the Navajo miners. She thought his answers were excuses for the industry, and she stormed at him. Nothing in her schoolwork prepared her for the impact of these broken Navajos' lives. She wept with indignation as she spoke of the mining conditions and their impact on the Navajos' health.

"As if they didn't have enough problems with tuberculosis and their drafty hogans which can't keep out the winter winds." She showed her father photos of the dry desolate huts.

The stories told her by these Navajos, delivered without recrimination, politicized Colleen.

When she graduated from high school, she won a scholarship to Berkeley to study microbiology. She accepted it quickly, although her mother tried to tempt her with Stanford.

"Wouldn't you like to join a sorority?"

"No, mother."

"But they have so many nice girls with your background."

"I won't have time for socializing," she said.

Colleen started in zoology, emphasized biochemistry, moved into virology, and never looked back. Five years later, she stood among the top students in the doctoral program.

PLANS FOR THE VIRUS

Colleen and Jonah argued about leaving Berkeley to take the job offer with Biosolz. Jonah regretted interrupting the doctoral program. Colleen needed money to complete her virus research. Jonah feared Colleen only wanted the prestige. But when the time came, they boarded the Lear jet and left the San Francisco Bay Area.

Regis joined them because he convinced his law partners there might be profits down the line. Jonah sought him out.

"I haven't thanked you for helping us," Jonah said. "I think you saved our lives by getting us out of Berkeley, and we would never have called Birnham or Biosolz."

"I thought Colleen might give me hell for butting in, like she used to," Regis said.

The pilot announced the approach to Raleigh-Durham airport where a limousine waited. The jet taxied close, and dropped the stairway. A trim gray-haired woman in a business suit walked to the plane and shook hands with Jonah and Regis. She embraced Colleen, then held her at arms length.

"I am Mary Henly. I'm the majority stockholder and Chief Executive Officer of Biosolz Corporation. I have heard about some of your difficulties. I promise you are safe now, and among friends.

Welcome to your new home."

Colleen collapsed in Mary's arms. "Thank God." The two women, arm in arm, heads together talking, led them to the limousine.

The bottle stolen from Regis' boat, with its solution of Colleen's virus, sat in the middle of a wooden table in a dimly lit room in Baghdad. Two people debriefed Mahmoud, after he arrived alone in Iraq with the bottle.

"Why do you ask about Waleed?" Mahmoud complained. "I did the work."

"Waleed led the team, but only you returned. Why?"

"Ah," Mahmoud grimaced. "We followed the man and his woman to a boat and found the bottle. Waleed told me to take it to the truck while he stayed to silence the man."

"What went wrong?"

"I ran to the truck. I heard shots. Waleed and Kahlil fell to the ground. Bullets hit the truck so I left to make sure the bottle got here. So don't call me coward." He glared, fingering a knife on his belt.

"I call you coward," Kahlid said. "You allowed my cousin to die."

Mahmoud's face darkened. The leader struck the table with his fist. "No fighting. Finish the story."

With his eyes on Kahlid, Mahmoud said, "During the shooting, I heard someone cry, 'Allah, Akbar.'"

"Other Muslims fired at you?" the leader asked.

Mahmoud shrugged away the question.

"Perhaps we have competition from the Iranians?" the leader said. "Anyway, it doesn't matter now. We have new orders. Another team will take the bottle back to the United States."

"Why the United States?" Mahmoud looked stunned. "We should take it to Israel."

"Who knows. The Council has ruled."

Mahmoud issued a string of epithets. As he left he spat over his shoulder at the leader.

"He is right," Kahlid said. "We should spread the virus in Israel."

Colleen and Jonah awoke feeling jet lag. From the bed, they

inspected their luxurious new quarters, glowing in the morning light.

"This looks very expensive," Colleen said, peering at the stained wood, high beamed ceilings, floor to ceiling glass, and green lawns and trees visible from every window.

Fleets of geese and ducks of all sizes and colors solemnly worked the manicured lawns, and slid into an occasional pond. After a knock, a cheerful young man arrived with fruit, rolls, and a pitcher of coffee. "If you need anything, just call."

They picked their way through exotic fruits and varieties of filled and glazed pastries. "What is that?" Colleen asked.

"Kiwi fruit, I think."

"Have you ever had a breakfast wake-up like this."

"Never. How come we don't know anything about each other?" Jonah asked.

"We've been busy. We only met the day you protected me from thugs and murderers. It started when you told Michael Bearing to 'shut up' and quit bullying me."

"You noticed that?"

"I noticed it. I loved it. And I never thanked you for it. Come here." She pulled his face against her breasts and kissed the top of his head for a while, rocking from side to side.

Jonah could hardly breathe, but he refused to move. Then she pulled herself down facing him, and wrapped her legs around him. She reached down, and when she had him properly aimed, fixed him with a severe gaze.

"If you ever need anything from me, promise me you will ask, or at least tell me what you are feeling. Promise me?" She kissed him. "If I can prevent it, I don't want you unhappy. Promise?"

"I promise."

With that, they went critical and melted down into a puddle. The rays of North Carolina sun bathed them in its morning heat.

They woke again at noon, showered, and slowly assembled themselves. They found their clothes brushed, pressed, and folded.

Mary Henly called. "Have you recovered?"

Colleen said, "More. We are reconstructed."

"Good. Would you like to join me for lunch?"

They followed Mary's directions and arrived at a large interior atrium garden, with walls overgrown by a flowering vine. Hors d'oeuvres and wines made up an elaborate table setting.

Before Colleen and Jonah arrived, Regis took Mary Henly aside.

"You know that I am a member of a small law firm?" Regis asked.

"Rosehill, Whitfield, and Jameson, out of San Francisco," Mary said quickly.

"My sister asked that I represent her, and my firm has taken an interest in her progress," Regis said.

"I am pleased that she will have dedicated legal advice," Mary said. "It will help us assure an equitable arrangement."

"I knew you wouldn't be concerned," Regis said. She's pretty smooth, he decided. They returned to the festive table.

When Jonah and Colleen arrived, Mary Henly again embraced Colleen and sat her in the chair next to Mary's.

She introduced everyone to two new people.

"This is Dr. Susan Trey and Dr. Ted Damkal, microbiologists from our staff." After the greetings, Mary turned to Jonah and spoke to the group.

"I am told we are here only because you took up Colleen's cause at several points? True?"

Jonah looked uncomfortable. "We had some very bad times which I don't want to talk about. I'm here quite by chance. Colleen and I want to apply research to the real world." Jonah hesitated. "I look forward to working with her. She is more enjoyable to talk to than anyone I know."

Colleen rose from her chair, went to Jonah, and kissed him on each cheek. She lifted her wine glass, and said, "This is my colleague who defended me and saved my life."

"Bravo," Mary Henly said as they applauded. "An auspicious beginning. Perhaps we could eat a bit and get to know each other better before we turn to the future."

Food was served, and the four biologists spent the afternoon sniffing around each other like young dogs establishing the pecking order, parceling out the territory, and scent-marking their specialty areas.

"Where did you do your graduate work?"

"Oh, Stanford?"

"You worked under Dreisser? What is he like?"

"Buford at USC? Oh, a good man, and a great teacher."

With their mentors identified, they inquired into each other's research interests.

Mary Henly gathered them together to close the session.

"We must become friends and confidantes, and we must trust

each other. We must do all this quickly." She turned especially to her two employees, Susan and Ted. "Otherwise, our new friends may not wish to share with us the information they have. If any of you feel uncertain, or find yourself distrustful of any of the others, please tell me, so that we can try to resolve problems early." She stopped, and shook her head slightly.

"I'm not just trying to be pleasant. A great deal of money will be invested in support of the information which you four must share. If our project is to be aborted, let it be early, rather than late. That will save us pain." She stopped and looked around.

"We agree, then?"

All nodded.

"Excellent. Go play my young friends. Dinner at six p.m. Breakfast, tomorrow, at seven a.m. Then we will get down to business."

Regis joined Colleen and Jonah at dinner in order to brief them. "Don't discuss the past situations at all. Focus on your product and how it affects human reproduction."

Jonah and Colleen nodded.

"Your job is to turn the information you reported to the Birnham Foundation into a patent, if possible. From the patent, we can extract profit."

"How would the ideal agreement with Biosolz read?" Jonah asked.

"Something like this: 'The compound or life form which interferes with human reproduction is to be described for purposes of propagation, development, and study under contract to Biosolz, Inc. Patent rights will be sought concurrently, in the name of Colleen Jameson.'"

Later, Jonah asked Mary Henly, "What if Colleen shares her product with you, and then finds that the effects are dangerous, perhaps catastrophic. What happens?"

"We'll stop using the product," Mary said, "and we'll warn anyone who could be harmed."

"What if you do not act in accord with the agreement?" Jonah asked.

"You can take us to court," Henly said. "The only real protection you have when you enter a contract is through the courts."

Colleen said, "I want your informal assurance that you will uphold the agreements we make."

"I promise you we will stand by our agreements," Mary said. "Also, we will discuss all agreements with your legal counsel, and he can review them for your further approval."

"Also, we will discuss all agreements with your legal counsel, and he can review them for your further approval."

Colleen turned to Regis.

"All right. I'd like you and Mary to work out the agreements, and then let us read them."

"We'll do a rough draft," Regis said.

Mary agreed. "First thing in the morning."

11

FIRST USE

The stolen virus sample went first to medical faculty members of the university hospital in Baghdad, Iraq. The physician who specialized in viral diseases spoke.

"We assume the material is a virus. We will turn the sample over to the production team at this time. They will first cut it into portions, and provide our political agents with samples for field testing."

"Where will it be tested?" a listener asked.

"In the United States, of course." The spokesman smiled. "The remainder will be the starter to grow more."

Four teams received small portions of the fluid in bottles labeled as a commercial cologne.

They traveled to the U.S. and took up positions around the United Nations building, the Pentagon, Wall Street, and the headquarters of the Jewish Defense League. They mingled with crowds, walked through buildings, visited restrooms, waited in crowded areas, and occasionally dispensed a small amount of spray.

The teams that distributed the aerosol sent a letter to the media:

"The first blow has fallen. The American military murderers have had their testicles cut. The United Nations' running dogs of corporate imperialism have also lost their inheritance. The capitalist gamblers of Wall Street should look to their futures, which they no longer have. Greatest Satans of America, reach down and feel for your nuts. Feel them for a last time, because they are gone. Let the war begin!"

ENVIRONMENTAL CLUBS

Wilson Hu washed up after the commute from the Wildness Club western office in Los Angeles to his Valley apartment. He had to chair an emergency meeting of the Wildness Club executive board in one hour.

The landscaped trees in the apartment complex allowed him a brief memory of his youth in Hawaii. How did I ever get stuck in this earthquake-plagued concrete mine-field, he wondered? Everything is suit-and-tie negotiations with lawyers and developers. How different from the tropical rainforests, volcanic eruptions, hurricanes, and tsunamis he grew up with. Visions of green foliage and flowers, and fields of wild orchids filled his memories.

Wilson sat down to return a call from someone named Regis Jameson. The name sounded familiar, if not Regis' voice.

"Are you any relation to the Berkeley student who discovered the new virus?" Wilson asked.

"My sister," Regis said. "I wish to bring some information regarding the new virus to your attention, but can only do so confidentially. Would you agree to keep our discussions private?"

"OK. I'll try to stop before I'm compromised," Wilson said.

"My sister Colleen wants to complete her research on the virus. She can't limit its spread since someone stole a quantity of the virus."

"I'm sorry to hear that," Wilson said.

Regis and Wilson discussed what might happen if the virus were developed in bulk. Regis continued.

"I'm Chair of the Western Division of the National Association of Riflemen. I wish we could try to identify some common environmental goals and combine our organizational efforts toward those goals."

"Mr. Jameson," Wilson said, "I can hardly believe the NAR would like to develop some common initiatives with the Wildness Club. Your members have been extremely critical of us in the past."

"I know, but I'd still like to start discussions. I'll have to see what I can sell to the members." Wilson sat wondering what he should do.

As one of the largest US environmental organizations, the Wildness Club had to respond to potential threats from population manipulation. Wilson's doorbell rang, and the Executive Committee members started arriving for the emergency meeting.

"Hi Polly, John. Sorry to roust you for another crisis."

Polly was up to speed, as always. "I'm worried about the Birnham Foundation. Do you have the newspaper report?"

Polly moaned as she read it. "Twenty thousand dollars. Most graduate students would kill for that. Not to mention the ego trip."

Wilson told John, "Some damned foundation put up a biannual prize. Hold on, Pete's here. He's with his new girl friend."

"I didn't get to know the last one." John said.

Wilson waved the newcomers to chairs around the kitchen table. "There's drinks and stuff. Sonja and Jules can't make it. Let's get started."

"OK. What's up?" Pete asked.

"A private foundation is trying to steer society toward their private agenda," Wilson told them. "The Birnham Foundation put up a $20,000 prize for anyone who comes up with a chemical, virus, or other agent that can interfere with human reproduction for several years."

"A chemical coitus interruptis," Pete said. "It sounds great. Long-term contraception."

"It says the agent has to be capable of retaining its effectiveness in an aqueous aerosol spray?" Polly asked.

"Right. What do you make of that?" Wilson asked.

"The bastards want something to spray over a large area," Polly said,

"Better yet," Pete said. "What's the hang-up, Polly?"

Polly glared at him. "Dammit, Pete. We gotta get you into a biology course."

"C'mon Polly. Just tell us."

"How far down our family tree will the spray be effective? Chimpanzees? Apes? Monkeys? Lemurs? Primates? Mammals at large? How many of these could be wiped out?" Polly looked at them angrily.

"Aw, they'd check something that important, wouldn't they?" He looked around at the silent group.

"No, they wouldn't," Wilson said.

"So what do we know about the Birnham Foundation?" Polly asked.

"We don't see any ideological position in the mission statement. They give a lot of latitude to the Board of Directors."

"Do we know anything about the directors?" John asked.

"Two have family planning backgrounds. The Chair is Moses Bradley, an early member of the spin-offs from the Population Institute, like the Commission on Growth."

"Sounds like a liberal," John said.

"I worried about that," Wilson said. "The other important player is named Jonathan Remmer. He's from Los Alamos and got future conscious. He thought up the prize and sold it to their directors."

"Any biologists among them?"

"No such luck," Wilson said. "The major trustee is named Marilee Birnham. All we know about her is that she is a musician. We have no information about what she cares about."

"So what'll we do?" Wilson asked.

"Well, let's start by asking our most prominent ecologists to write letters of warning about possible negative effects on other organisms. We can talk to the American Association for the Advancement of Science, for openers."

"OK," and I'll get it into the next issue of *Wildness* magazine. Wilson, you and Polly go talk to the Foundation Board?"

"Sure."

"Is there anybody to boycott?" John asked.

"No one so far," Wilson said. "The elder Birnham inherited his money. There isn't a corporation behind the Foundation. We're stuck with persuasion, influence, logic, and sweet reason."

"If that's true we're doomed." Polly shivered. "Our near relatives on the evolutionary tree may be in big trouble."

"We're finished, then?" Wilson asked.

"No," Polly burst out. "We have to do more than this. Call the Conservation Chair in the state where the foundation has its headquarters. Get them to condemn the prize. Then we can call in the Legal Defense Fund."

Wilson said, "And what do we ask them to do?"

"Get an injunction against offering the prize until the effects have been studied. We should demand an environmental impact study," Polly said. "Whatever will stop the award." There was a pause.

"Are we sure all these bad things are going to happen?" John asked.

"I can try to call in the Legal Defense Fund, but I always catch hell," Wilson said.

"From who?" Polly demanded. "It's important."

"The national officers don't like it when we reach for the law," Wilson said. "They argue that it is too confrontational."

"You and those national officers all need training in what happens if you keep fooling around with the natural world," she said. She started to say more, then grabbed her purse and stormed out.

"I guess we're through," Wilson said. That's how Earth Prime!

got started. Polly wants to work on fundamentals like human overpopulation, and the Club isn't ready.

A STUDENT FAILURE IS A TEACHER FAILURE

Dr. Jared Shevlin looked over the last of Colleen Jameson's notes which Michael Bearing gave him. Pity about Bearing, he thought. Should have been more careful. Shevlin asked his secretary to telephone the Birnham Foundation.

"To whom am I speaking, please?" Shevlin asked.

"Moses Tyler, President of the Board. What can I do for you?"

"This is Professor Jared Shevlin, Head of the Virology Department at Berkeley. I've lost track of a student of mine who may have applied for your prize. Her name is Colleen Jameson."

"Yes. She did apply for the prize. Her most recent forwarding address is Biosolz Corporation."

"In the Research Quadrangle?" Shevlin asked.

"Yes, in North Carolina. Is there a problem?"

"Oh, nothing I should let bother me. I served as the senior professor for the work she submitted to you. I thought we prepared it for joint publication," Shevlin said.

"Is that right? Will you be claiming priority of authorship?"

"I'd rather talk to her about a joint submission to your group. I'll call her and see how she feels about that. She's young and a bit headstrong."

"I can almost remember those days." Moses Tyler laughed.

"Thanks. I'll get back to you." Shevlin broke off.

My God, she got a job, thought Shevlin. It sounds like they have taken her prize application seriously. He wondered where she stored the virus samples. After some thought, he arranged for an appointment with the Senior Bishop of the Holy Father Church in San Francisco.

"How may I serve, Dr. Shevlin?" The Bishop appeared to listen very carefully.

I'll bet he hasn't talked to many professors of science departments before, Shevlin thought. He reviewed his version of Colleen's attempts to take advantage of the discovery of the virus. He described in detail how the application of the virus brought about involuntary sterility in males.

"Involuntary sterility?" The Bishop expressed great dismay.

Shevlin said, "I thought the Holy Father Church may wish to organize an inquiry and perhaps impose sanctions upon Ms.

Jameson to express the church's disapproval. I had no connection with her research and consider her line of inquiry unethical. Of course, I don't wish to make a public outcry, what with the University and my official position, you understand."

"Of course," the Bishop said. After Shevlin left, the Bishop asked an aide to prepare a full report on The Jameson Matter that he could forward to his superiors. He certainly wanted to alert them to any serious attack upon the doctrines of the Holy Father Church.

Last, Dr. Jared Shevlin called Biosolz Corporation.

"This is Mary Henly. How can I help you, Dr. Shevlin?"

"Well, I'm pleased to have caught up with my errant student, Colleen Jameson. She disappeared rather suddenly, and I hope that she is all right."

"She's just fine, and working hard," Henly said.

"Good. Her departure left some loose ends with her doctoral work. For someone so young, she may have had some promise. Its hard to know since she abandoned everything."

"Did you want to give her a message?" Henly asked.

"Better not, lest she think I'm upset with her. By the way, If you need any assistance with your viral development program, just let me know. I'd be glad to help clean up any deficiencies in Colleen's work."

"Thank you, Dr. Shevlin. I'll keep that in mind." So he's offering to help clean up deficiencies in Colleen's work? Mary wondered. For a stiff consulting fee, I'll bet. He sounds like a pharmaceutical detail man, ready to say anything to make a sale. She decided not to mention the call to Colleen.

12

GO/NO GO

Colleen woke Jonah with her tongue in his ear. "Wake up, dammit, Jonah. How can you sleep while I'm worried?"

"I can't." Surrounded by Colleen's warmth, he nuzzled her throat.

"Listen to me, Jonah. Are we about to join unscrupulous capitalists?"

"That is a redundant tautology." He kissed her on the breasts.

"Uh huh. But how do we get out?"

"We don't," Jonah said, continuing to kiss her.

"But why do we demand so many conditions?"

"To make them think we believe in them," Jonah said, positioning himself more advantageously.

She grabbed him by his ears, and held his head at a distance. "Pay attention, you!"

"I'm paying close attention." He nuzzled his way back between her breasts.

"And you don't believe them?"

"No. I expect them to screw us. Here, I'll show you."

"But don't you feel like a hypocrite?" she asked.

"No. I really enjoy you," Jonah said.

At breakfast, Colleen, Jonah, and Regis joined Mary, Susan Trey, Ted Damkal, and a new face.

"For our newcomers, this is Sean Levee," Mary said. "We will record our discussions so that we can resolve any differences of opinion later, if you don't object."

Sean appeared thirty five, tall and slender, with narrow features. He dressed in a business suit, with carefully matched grays and charcoal. Colleen looked him over coolly but then returned his warm smile.

Mary continued. "Sean helps with administrative matters. He schedules my appointments, and my travel, and is usually able to second guess me if you need a quick opinion on something."

"I trust him completely," she said. "I guess he doesn't quite trust me yet, since he won't travel with me. We shall see who wins." Sean smiled at them without embarrassment.

"Sean dear, please give them your twenty-four hour number," Mary said. Sean did so.

"Don't hesitate to call me anytime, even at night, or I'll be blamed for not solving a problem before it happens. I'm being paid well. Anyway, I fall asleep easily."

"He certainly does," Mary said. "Now, what do we need to get started? Jonah, what are your concerns right this moment?"

Why did she start with me? Jonah wondered. "From past events, we worry that Colleen's research may be so important that groups we don't even know will kill to learn more about her findings. I keep a pistol under my pillow, and I'd like somebody to convince me that we're safe in our beds, and in these rooms."

"Biosolz sounded perfect, with its perimeter fence, and controlled access," Regis said. "I hope this means that the residential, business, and laboratory facility are enclosed and safe from outsiders."

"Indeed they are," Mary said, "but you remind us that one can't work well while looking fearfully over one's shoulder. Sean, would you make the arrangements? We'll take you all on a tour of our security system right away."

Mary turned to Colleen. "What agreements do we need before we get into the technical production problems? Colleen? Regis?"

"I want to be shielded from political questions about what we are doing," Colleen said. "I want to concentrate on the work, and the research questions which drove my project at Berkeley."

"Of course. You want a research atmosphere," Mary said.

"I'm not as optimistic as Colleen," Jonah said. I don't think we'll ever be allowed back into a doctoral program. There seems to be a trail of violence behind us. Therefore, I am more concerned that we achieve financial independence. I think that we must emerge financially secure from our time with Biosolz or we will lose what we gained from our doctoral work up to now."

"I fear I agree with you," Mary Henly said.

Jonah continued. "Therefore, I'd like Regis to arrange for our financial security, assuming that Colleen's findings are that saleable."

"Speaking for the company, I certainly want to make a profit, so I hope we can move forward as quickly as possible," Mary said.

"In that vein, Colleen, are you ready to tell us the nature of the agent you have identified? Is it a virus? Chemical compound? Bacterium?"

Colleen sat quietly with her head down for several seconds. "I must not be as ready as I thought. I find myself filled with garbled, irrational fears."

"Can you tell us about them?" Mary asked.

"I can't fathom the violence that surrounded us. Will there be more? From where will it come? I don't feel safe enough to go forward. I still wake up frightened. Will things really improve? Maybe the review of security arrangements will help."

"Fair enough," Mary said, "Sean and I will arrange for an immediate review of security while we tour the grounds. We'll call you when we are ready."

In Atlanta, the Deputy Director of the Center for Disease

Containment stormed into the Director's office. A West Coast friend faxed him a story from the *Bay Area Chronicle*. "Listen to this."

"Highly placed government sources reported that teams from the Center for Disease Containment investigated allegations that employees of the Jewish Defense League and of Wall Street were sprayed with infectious material by persons unknown. The sprayed material may prevent exposed males from fathering children. A CDC spokesman denied the rumor, but said they routinely conduct studies of new viruses. Anyone who thinks they were exposed should call the CDC."

"What the hell," the Director said. "Who leaked this?"

"Probably the Secretary of Health himself. He prefers a bad press to no press," the Deputy Director said.

"Yeah, and that's not all. Those *Chronicle* SOBs didn't even consider it a page one scoop. OK, what's the rest?"

The deputy handed him a copy of the *Post*, dated one day later.

"Members of the activist Ottoman Committee told reporters today that the announcement which said Jewish Defense League members had been sprayed was a lie, spread to provoke enmity between groups. They said that members of the JDL refused to give sperm samples because they are impotent, and the JDL wants to conceal this fact. 'They have no balls,' claimed spokesmen for the Ottoman Committee."

"Jewish Defense League members hotly denied this claim, and told the *Chronicle*, 'We'll show you who has no balls.' When asked what this meant, the JDL members refused further comment."

A few days after the FBI received information about the virus, they requested a meeting with the CDC Director.

The FBI Agent demanded, "Tell me whatever you can about the virus."

"We don't know much at the moment. We are tooling up to see what kind of microbe we are up against," the CDC Director said.

"Is it deadly? Is it infectious?" the FBI Agent asked. "How long does the sperm immobility last? Is it permanent?"

"That's what we are trying to find out, what kind of virus it is. The answers to your questions are two years away. May I ask you what microbiology training you have had?"

"I have an MBA from Georgetown," the FBI agent told them,

"I thought so," the CDC Director said. "Next time send a technician to ask questions. You should leave and come back when we can find some answers."

Three weeks later, the CDC Director held a staff meeting.

"OK, here is what we think, so far, and it is not for release to the press. The virus is presumed to be air or water borne, and has an incubation period of one to three days after initial contact with a male. Males feel a slight pain in the testicles as the only noticeable symptom of infection. With most infected men, functional sperm disappear within a few hours or days. In a small percentage of the men examined, we observed sperm only partially immobilized. Until our public relations people make some decisions about policy, this information should be treated as confidential."

Colleen and Jonah fought all evening in their Biosolz quarters.

"Jonah, these are the only friends we have. We have to trust them."

"I agree," Jonah said. "And Regis should return to his law practice, with an open line to us."

"He doesn't want to leave, yet."

Jonah said, "I want somebody on the outside, not under the control of the company."

"OK. But you agree that Biosolz is our last hope?"

"Yes, actually I do," Jonah said.

Regis balked at leaving. "I can make sure you are safe here. They don't need me in the San Francisco office."

Jonah said, "I'm more concerned about what is going on outside. Here, we'll be limited to what the company tells us. I want to be able to call in the cavalry."

Finally, Regis returned to San Francisco.

Colleen and Jonah had one more struggle with Mary.

"Who will control production, once we start?" Colleen asked.

"You will always know more than the rest of us about the agent. You'll be in the lab." Mary said.

"What job has highest priority?" Colleen asked.

"Producing larger quantities of the agent than we now have," Mary said.

Colleen asked, "Who makes the decisions about distribution?"

Mary said, "You and Jonah decide. You brought it to us."

"I want Jonah part of all decisions in the future."

"Certainly," Mary said. "He can check your thinking."

Regis approved final contract papers for Colleen and Jonah's signature. As they signed the last sheet, Mary turned to Colleen.

"This is the last bit of information we need. What is the active agent? A chemical compound? Or a virus?"

Colleen took a deep breath. They could not turn back once she told them. "Virus."

Mary Henley, PhD in Pharmacology, leaned back in her chair, and looked at the ceiling.

"Oh God." Then she lowered her gaze to the others and said, "You know, I really hoped it would be a compound. No self-replication or mutations to worry about. Then we could leave everything to the biochemists." Mary fell silent for a bit and laughed wryly.

"We'll have to freeze semen samples from all the technical males." Jonah liked Mary better when the scientist in her peeked through the business shell.

Finally Mary shook herself. "So be it. Let's get to work."

For the rest of the day, Susan walked Colleen and Jonah through a well equipped virology laboratory on the company premises. Susan described the newest high technology equipment which would permit the introduction and isolation of a virus strain, growth of that strain, and preparation and storage of batch lots. Then Susan asked, "So where do we get our starter sample of the virus?"

Jonah said, "I'll go get it." Jonah wondered why he suddenly didn't want to tell where they hid the virus.

"Mary might object," Susan warned. "You agreed to stay on the premises during the development work."

Mary called a special meeting that night. "What is this about Jonah leaving the premises? You signed agreements that both of you would stay here until we finish the development effort. Work straight through, no vacations."

"I just prefer that Jonah pick it up," Colleen said. She began to feel uneasy.

"But can't Ted or Susan get it more easily?" Mary asked. "I assume Jonah may be sought by the police." Her voice rose with impatience.

"I really want Jonah to be responsible for it," Colleen said. Why am I being this stubborn? she wondered.

Mary drew herself up. "Are you sure about this?"

"Yes." Colleen wasn't sure, but it felt good to fight Mary.

"If that is your decision, you have just broken our new agreement," Mary said. "I will provide ground transportation for you tomorrow morning to Raleigh Durham Airport, with one-way tickets to whatever destination you select."

She dismissed them, and left the room.

13

PRODUCTION

Once again, Colleen and Jonah fought all night. "We have no protection from Mary's stealing everything from us," Colleen said.

"And we have no other place to go, nor money to get there," Jonah said.

"You would rather stay here?"

"Yes," Jonah said.

"So you'll tell Mary we've capitulated?" Colleen asked. "I'm embarrassed." Colleen tried to cling to her rage at Mary, but her fear of leaving took over.

They told Mary together.

Colleen, Jonah and Susan got the lab ready, while Ted flew to California to retrieve the virus sample.

While they waited, Colleen dreamed troubled scenes of the future and couldn't understand why. Her own brother applied a lawyer's blessing to the plans. Mary Henly provided managerial oversight and financial backing. Jonah gave moral support, although he would have called it immoral support. Colleen hadn't the will to withdraw from the commitments made so far.

She couldn't walk away from her comfortable quarters, without money, potentially a fugitive from the law, perhaps followed by a new set of pursuers, with no place to go. In bed, she moved closer to Jonah for reassurance. He put his arms around her and said, "Don't worry, everything will get worse." She moaned and went to sleep.

When Ted returned, they took the samples to the Lab and ran initial viability checks. It appeared they had live specimens, or more properly, reproducible molecules.

Mary called for a party that night. They all got drunk. Colleen passed out muttering, "Allah Ach'bar . . . I am destroyer of worlds. Oppenheimer was right."

Next morning, Mary Henly threw a newspaper in front of Jonah during breakfast. He read the article that described the political group's application of the spray to people around Washington, DC.

"It probably means that the Center for Disease Containment has now entered the action. Your Iraqi friends released the virus they took from you," Mary said.

"I still have the feeling we should notify the government," Colleen said. "Isn't there some way to tell them?"

Mary Henly said, tartly, "I'm going to pretend that I didn't hear that, unless you want to spend a long time in jail. You must distance yourself from anything that ever happened to you since Michael Bearing's death. Jonah, would you two please spend some time discussing the ramifications of this?" Jonah nodded.

Mary continued. "First, the University legal staff and ethics committee would grill you to make sure that you, not they, were the cause of any problems that might emerge. Second, the Berkeley Police still have some unresolved murders. They would want to talk to you for a while."

"Then there's the California Harbor Department which might want to know about illegal activities involving your boat. The FBI and CDC will want to know about any production or transportation of unregistered biological agents. The CIA will want to ask about dissemination of biologicals to overseas political groups."

"But we did it under duress," Colleen said.

"Oh did you, now? Do you want to try and prove that to an eager prosecutor who must find someone to scapegoat and bring someone to trial?" Mary shook her head at the two young people. "Think, people, think."

Colleen shuddered. Jonah wondered how Mary had learned so much of their past. "So what do you recommend we do?"

Mary said, "Work as hard as we can to produce a large, stable, replicable, virus culture so that we can develop a vaccine."

"I like that," Colleen said. "It makes it easier for me to dive into work. I don't like this political use of the spray." Poor dear, thought Jonah. Colleen had the guilties, and he couldn't help her. By contrast, Jonah felt such anger at the damage many humans inflicted on the world's environment that he didn't care who used the virus.

PRODUCTION TESTING

At the next business meeting, Mary asked about the production process. "How is it coming?"

"We're doing fine," Colleen said. "The first batch is nearly ready for storage. How about the vaccine development? Any progress?"

"No," Mary said. "Our colleagues in Chicago fear antigenic drift. It's going to be a long tough problem. We may have to bring in some other labs."

Jonah awoke to find Colleen fitting him with a condom.

He opened one eye, and announced, "A girl once told me 'You no kiss me because you love me. You kiss me because you want to do me something.'"

Colleen stopped what she was doing. "What?"

Jonah explained, "A girl told me that once in San Antonio. She communicated very effectively."

"Well she certainly did," Colleen said. "Now let me try to communicate just as effectively. Since you distrust my motives in fondling you, I'll stop if you insist."

"I didn't mean for you to stop," Jonah said.

"Don't interrupt," Colleen said. "As I was saying, tell me where you want me to get my sperm sample for todays lab work."

She rubbed him some more. "Would you like us to nominate you to the IRS as a 'provider,' or 'source?' Perhaps you would like to have those tax advantages later, that is, if you can find it in you to perform. But maybe it is too much trouble?"

She stopped fondling him.

"Don't stop," Jonah said. "I just wondered . . ."

"Uh huh. Perhaps you should wonder whether you are interested in informal power? Or prestige? Would you like us to call you 'Head Ejaculator,' or some other finger of speech?"

Colleen stopped rubbing again.

"No, I . . ." She cut him off again.

"Let me know, now, because, if you don't want to be a specimen provider, I need to leave you so as to search out the services of someone else. Who would you recommend? Sean Levee?"

Jonah frowned.

"Or, imagine, my dear, if he were as picky as you, and rejected me, then I would have to go and ask for the assistance of Ted

Damkal. After that, I'll have to approach the stable of drivers and male house staff. What do you suggest?"

"Well, I just meant that you shouldn't do that to a person without kissing them. Uh, him. Me."

"Oh, poor dear." Colleen kissed him, fondled him some more, and soon had her specimen for the day's work. After all, she had to make frequent quality control checks on the rapidity with which the virus strain growing in the laboratory would kill live sperm.

In the lab, Colleen asked Ted Damkal, "Where is yesterday's batch?"

Ted looked startled. "I thought you wanted it filtered. I moved it to storage."

"Ah," Colleen said. "You left some starter?"

"Sure. Vat Three," he said.

"How much do we have now?"

"I don't even know. I've been running back and forth with the nutrients."

"I'll ask Susan," Colleen said, and left.

Mary Henly found Colleen muffled in full battle gear: canvas bootees; pull-on high rubber fishermen's boots; canvas lab pants with overall straps; canvas pull-over jacket, wrap-around style to protect from spills; surgeon's cap to contain her hair; double disk respirator; and canvas hood with front visor panel hanging well down over her jacket. Finally, an air tube ran to the hood to keep positive pressure and fresh, filtered air flowing to her in case of a major spill. That would give her time enough to get out and decontaminate herself in stages. All this to prevent no known effects of the virus upon women.

Colleen had her hands in an air-proof work station called a glove box. It had a transparent front panel and openings where she could insert her arms into the long autopsy gloves sealed to the edge of the access ports. She introduced specimens and needed equipment through an air lock on the side. She operated the equipment and manipulated specimens while they remained isolated within the box.

Mary watched Colleen's movements in the glove box. "What's up?" she asked Colleen.

"I tagged some of the substrate nutrients with a radionuclide. I'm taking a sample for assay to see where it goes during sperm metabolism."

"How do you follow them without radiation meters?" Mary asked.

"Jonah and Ted cranked up a radioassay counter. Go see it. It's ingenious. Ted really expedited this. Will you thank him for me?"

“Indeed, I will.” Ted’s contribution pleased Mary.

“How many batches of filtrate do we have so far?” Colleen asked.

“We have three batches of two hundred liters each in the cold tanks,” Mary said.

“Sold anything yet?”

“No, but we have some nibbles. We will soon.”

“Good price?” Colleen asked.

“Very. But, our early expenses are high.”

“Sure.” Colleen turned back to her task, dismissing Mary.

14

FIRST CORPORATE PROFITS

While Colleen and the virus growth team enlarged the yields of the production labs, Mary Henley began to draw off the virus into larger storage tanks. She made preparations for the first batch sale. She asked Northby’s International Auction House to sell it for her on commission, to a limited audience. Mary handled the advertisements, a letter to a few dozen of the heads of the most oil-wealthy families in the middle-east region. The auction house sold the one-ton batch of virus disguised as an art object, its true nature concealed.

The auctioneer now offered the last art item before Mary Henly’s sale.

“Ladies and gentlemen,” the auctioneer intoned, “we now offer item twenty three, a newly found keyboard score by Domenico Scarlatti. It is described in your directory. Music lovers are excited by several new ornamentations, not seen before in this artist’s hand. Who will start the bid?”

A timid harpsichordist in the front row raised his hand. “Ten pounds.”

No one else came forth.

“Sold to Mr. Smythe.”

“And now we turn to the last item in today’s auction. We have a manuscript by a lesser known writer, Jesu Lesistratus, from the

18th Century, not available for inspection. It arrived too late for description." The sophisticated audience took a "ho-hum" position and started to leave.

"What am I bid on this item?"

"Five thousand pounds."

"Five thousand pounds? Yes. Would anyone care to raise the bid?"

A hand flickered toward the back.

"Sir? Ah, yes. Ten thousand pounds. Do we . . ."

"Fifty thousand pounds."

By this time, the regular art connoisseurs of Northby's knew something was up. A ripple of excitement went through their ranks. One of the wealthier among them stood and announced:

"Fifty five thousand pounds." He smiled all around to applause from his friends and wondered what he had bought.

"Do I have another bid, ladies and . . ."

A hand waved from the phone bank where anxious agents conferred with clients on other continents.

"One hundred thousand pounds . . ."

The jet-set adventurer suddenly sat down, realizing he wasn't in the right league.

"Five hundred thousand pounds," a voice toward the back said.

"One million," a swathed sheik said, without looking up.

The Northby auctioneer tried to maintain his composure. "Certainly, one million pounds. Do we have a higher bid?"

"Five million," nodded another middle eastern magnate.

"Ten million," a third shouted.

"My last offer: eleven million pounds!" another said.

The Northby agent's knees buckled and he was fanned where he lay alongside the rostrum.

The auctioneer's supervisor appeared, and picked up the gavel.

"We have eleven million pounds, going once . . ."

"Fifteen million."

Northby: "Going once for fifteen million . . ."

"Twenty million pounds."

"Twenty million. Right. Are there further bids? Do we close the bidding?" He waited an agonizing four seconds. "Last chance to improve the bidding on the Jesu Lesistratus manuscript?"

The jet-setters riffled through their directories, wondering how to spell Lesistratus.

"Going once, going twice, sold to the gentleman for twenty million pounds." With this the second auctioneer slid down alongside his colleague.

By the time Northby's controller collected the money and closed the transaction, the auctioneers regained their composure and headed for the bar. Northby's took its commission and invited all to return. Mary Henly's first sale and profit-taking had been consummated

The Iranian petroleum magnate who won the bid exchanged twenty million British pounds for ownership of one long ton of Colleen Jameson's or rather Biosolz' virus solution, ready to spray. It was the first commercial batch to be offered and would likely cause the other bidders present to seek their own supply.

The buyer, an Iranian Shia Moslem, pumped petroleum from the largest anticlinal oil field in the world, a huge underground structure that forced oil up the drill holes through its domed surface into Iran's tanks.

The buyer packed the five 55-gallon drums of virus into his jet. As he watched the drums loaded, he tried to calculate how many barrels of this size he had to fill with crude oil and send to the West to earn 20 million pounds. He lost track at something over a million barrels, little more than a drop in a dome. Then he and his jet flew off to Teheran, Iran. Now, he had to find a spray plane.

The Berkeley Police Chief hung up the phone following a call from the FBI in Washington, D.C. He asked his senior investigator to join him.

The Chief remembered the days when the city of Berkeley won a national award as "Best Small Community Police Department," and wished for their return. The investigator came through the door.

"Alex, help me with a case," the Chief said. "You remember the bodies we found in the Berkeley Marina a couple of months ago?"

"The three foreigners?"

"Yeah. Whatever happened to that investigation?"

"The men held Iraqi passports. The Iraqi foreign students at Multinational House claimed they didn't know them."

"You think they told the truth?"

"Yeah. They didn't seem to care about them. The two students who went to the morgue said they looked lower class."

"How could they tell?"

"Damned if I know. Two carried small stilettos, made in Iraq."

"All of them shot?" the Chief asked.

"Yep. With a dozen or so .22 slugs, probably all from the same weapon. We never found it."

"No witnesses?"

"No. It happened early in the morning, maybe three a.m. There was a wind which made it hard to hear. A couple of people a quarter-of-a-mile away said they heard some pops, maybe shots. They didn't pay any attention. The security patrol found the bodies when they made their five a.m. rounds."

"So what do you think?" the Chief asked.

"We wrote it up as a possible drug involved incident. The Marina had a couple of attempted landings, from small craft, over the years."

"What about the consulate in San Francisco?"

"According to them, the dead men entered the US as visitors, six days before they died. Like they had a job to do."

"So you closed the case?"

"Yeah. You want to reopen it?"

"No." He paused.

"How about the killing, around that time, near the University, Channing Way?"

"Yeah. A student, knifed."

"What did you get there?"

"Again, nothing. Someone killed the victim in his apartment at two in the morning. A neighbor heard noises and called in. We sent a car out, but the caller didn't see anything. No witnesses. No fingerprints."

"What did his classmates say?"

"Nobody in his graduate department had any idea who would kill him. Of course, he was a pharmacologist."

"Pharmacist?"

"Nope. I made the same mistake. The students gave me hell, and let us know a pharmacologist is a doctor, a top-notch physiological chemist who studies drug actions in humans. A pharmacist counts out pills."

The chief showed some interest. "Drugs again?"

"Maybe. The lab had a lot of stuff around. The students said a person might be able to support a habit, but they doubted the victim was involved."

"One other incident. A guy checked into the Berkeley hospital emergency room about the same time as the knifing?"

"Yeah, cut his arm bad. Kept cussing some woman and raving about God and justice. He wouldn't tell the hospital what happened, 'just an accident.'"

"That's it?" the Chief asked.

"Yeah."

"Thanks Alex."

The Chief called the FBI agent back. "We can't place your

woman, or any one at all near the crimes. What do you have?"

"I guess nothing," the agent said. "We had a rumor from an Iraqi informant that involved a woman. We hoped you knew her."

"Sorry." He hung up and abandoned the inquiry.

15

BIRNHAM SCREW-UP?

Moses Tyler, Chair of the Birnham Foundation Board, regretted that he recommended Colleen Jameson to Biosolz Corporation. Ever since the Birnham Board members offered the prize, they anticipated public criticism, and now Marilee Birnham had called to demand a board meeting.

Chairman Tyler and Jonathan Remmer prepared for the meeting at a hotel near Chicago's O'Hare Airport.

"Marilee Birnham sounded angry when I called her," Remmer said.

"What's new," Moses grumped. "She demanded a meeting. Immediately. We'll be lucky to get a quorum on this short notice."

Jonathan Remmer, the intellectual leader of the Foundation Board, looked worried. Moses continued.

"I'd appreciate it if you would take an active role this afternoon. Marilee listens to you."

"What's the mission?" Remmer asked.

"Since Marilee usually ignores these meetings, and we've had some bad press lately, she may be coming to fire me or withdraw the Birnham Prize."

"I'll damned well fight that. You know I'll support you, Moses. Hang in there, and call on me whenever you want."

"I appreciate that."

Moses Tyler and Jonathan Remmer were deeply worried about world population growth, and felt they should do more than wring their hands. However, Moses feared that most members joined the Board only to seek money for their own projects. He could not predict how they would act in a fight.

The relatively small amount of money offered as the Birnham

Foundation Prize produced highly visible results. Since Marilee Birnham seldom came to Foundation meetings, no one knew how she would react. Moses expected the worst.

He watched Remmer hurry to Ms. Birnham as she entered the conference room. Moses always assumed Remmer was in love with her. Remmer now kissed her cheek and helped with her coat.

A striking woman in her mid-fifties, she dressed in black. She had dark hair starting to streak, and a willowy figure. Educated in the best of the women's colleges, she completed her debut as a pianist before she married elderly Thadeus Birnham. Upon the death of her husband, she was left a wealthy patron of the arts and trustee of the Birnham Foundation. Jonathan Remmer pursued her.

"Have you decided where we should travel? I have a cruise planned for New Zealand. Come with me."

"I'm not going with you, Jonathan. A trip like that would scandalize my daughters, even at their age."

"Then marry me, and we can go as newlyweds on our honeymoon," he said.

"Jonathan, be serious. We'd better sit down." She walked him to the conference table.

A seventh board member arrived and provided a quorum. Moses Tyler gaveled the meeting to begin.

"Members of the Board, I thank you for coming on short notice. We especially welcome Trustee Marilee Birnham who asked for a special meeting." He turned and nodded, almost a bow, to her.

"Thank you Moses," she said, mildly. "I hope we can discuss this group's latest screw-up, if you will excuse me for speaking plainly. I refer of course to the Birnham Prize. Which one of you idiots thought up this one?" Ms. Birnham lost her softness when surrounded by fawning seekers of her wealth.

"Of course, you know I proposed it, Marilee," Jonathan Remmer announced. "I always come up with ideas which upset you. Don't be too angry until you've heard our side."

"I've asked Jonathan Remmer to get us started," Moses said.

"Yes. I move a vote of confidence for Chairman Moses Tyler, and urge him to keep working on the prize and population plan previously approved by the Board." Someone seconded "for discussion."

Moses began. "Ms. Birnham reports that the virus released in Washington, D.C. may have originated from one of our applicants for the Birnham Prize."

Two more board members arrived. Moses waved them to chairs and continued.

"As you recall, our population analysis group told us the earth can support less than 20 percent of our current population."

Tim Gedland, a Canadian economist new to the board, raised his hand. "Would you briefly review the data for us? Also, where the numbers came from?"

Chairman Tyler summarized findings of the World Watch Institute and a Berkeley interdisciplinary Population and Resources Group. He projected the loss of oil for energy and agricultural chemicals and stated that more than five times as many people are alive on earth than can be supported over time.

Moses did not like to present numerical data verbally without charts and turned to Marilee. "All right so far?"

"Don't you patronize me, Moses Tyler. I may be a woman, but I can follow statistics. Get on with what you are trying to tell us."

Economist Gedland jumped in. "There are plenty of fuels. If petroleum becomes too expensive, coal, natural gas, and nuclear energy can be substituted "

"No," Jonathan Remmer said, "The world supply of petroleum is nearly gone. From present trends, the population of China will double in 35 years. Just the coal they burn will produce enough pollutants to kill us all and further damage the ecosystem."

"Most economists don't take those projections seriously. Substitution of other fuels is likely."

"Mr. Gedland, your economic models are flawed. They fail to account for resource depletion. You wrongly assume that substitutes can be found for any dwindling raw material. You cut down trees and call them assets. You ignore health costs of the auto industry and decommissioning costs of nuclear energy. Economists' predictions have proved to be no more accurate than flipping a coin. I recommend to you the biophysics literature that describes the limitations on how much life the earth can support."

"Why must the Chinese burn the coal?" Ms. Birnham asked.

"The simple answer is that they have the coal reserves and they also have a fast-growing population clamoring for fuel," Remmer said.

"What are the alternatives?" Birnham asked.

"We should be developing large sustainable solar systems—probably photovoltaic—with our remaining fossil fuel supplies, " Remmer said.

"So why don't we?" Ms. Birnham asked.

"Our largest multinational corporations are oil companies," Remmer said. "They seem determined to wring out the last profits from our dwindling petroleum supplies."

"Why doesn't the government support photovoltaic solar research?" Ms. Birnham persisted.

"President Carter provided subsidies, but Reagan and Bush reassigned them to fossil fuel and nuclear industries," Remmer said. So oil companies cling to their subsidized profits until all petroleum is gone."

Moses Tyler broke in urgently. "It appeared that any plan which could reduce population would be socially unacceptable, and any socially acceptable plan wouldn't reduce population. Most members of the Board found this impasse disturbing."

"And this brought you to the Birnham Prize?" Marilee asked.

The Chairman nodded. "Our Jonathan Remmer worked on the Manhattan Project and concluded that a determined group of scientists can accomplish almost anything." He turned to Remmer. "Jonathan, please finish this story."

Remmer rose to his feet. "The crash project to develop nuclear bombs ended World War II. Yet, three weeks before their use, the designers did not know whether plutonium atomic bombs would work. They only had theoretical predictions to guide them."

Remmer continued. "A band of young physicists achieved this astonishing design feat. So it seemed to me that a self-selected group of ego motivated graduate students working on population control might accomplish the impossible, just as they did at Los Alamos in the 1940s."

"But what about the global impact of a population control agent?" Marilee asked.

"We anticipate political and social disruption," Remmer said. "But burning large quantities of coal will disrupt human society even more, and nuclear wastes may well wipe out all life."

"So you placed the advertisement," Marilee said. "Moses told me of a student who recently discovered a virus which disrupts reproduction. She is one of our applicants?"

"Yes," Remmer said.

"The good news is that her work is creating much interest," Moses Tyler said. "The virus has been actively sought by other governments."

"Yes," Marilee said. "My own inquiries told me that at least two groups of terrorists, plus our own CIA, the Mossad of Israel, and others are currently evaluating possible use of the virus."

"I was told that the Peoples' Republic of China approached the White House to buy samples of the virus," Jonathan Remmer said. "They didn't even ask the price."

"Shouldn't we withdraw the prize to avoid future liability?" Marilee asked.

"I'm not sure we could escape that easily," Remmer said. "The federal Center for Disease Containment investigates reports of new pathogenic viruses. If they find them, they will diagnosis the virus type, and try to produce vaccines. The Center is now checking for possible infections in federal agencies."

"No loss there," Marilee said. "So what can we do?"

"Our lawyers recommend that we say nothing about our program or our intentions regarding the Birnham Prize," Moses Tyler answered.

"Of course," Ms. Birnham said. "But I want to know what will happen in the future, and how we protect the Foundation from undesirable political backlash, not to mention angry litigious victims."

She suddenly threw up her hands in dismay, her voice rising. "All right, you fools. Now do you see why I don't trust the Directors of this Board? You started something which got out of control. Now my last question. Why didn't you would-be radicals approach an organization that knows what it is doing, like the Population Council? They at least have prominent corporate leaders who know how to manage. Well?" She looked around.

Moses Tyler banged the gavel, his face crimson. "I wish to speak to that comment at some length."

Marilee Birnham hid a smile. Finally, she observed to herself, I smoked him out.

Tyler took a deep breath and began with restraint.

"In the 1950s, foundations including Rockefeller and Ford learned how to spend our tax money to tell us what to do. They established the Population Crisis Committee and The Population Council. They used the Planned Parenthood Federation to disseminate policies to the community. Corporations established study centers at several universities including Harvard, Columbia, Michigan, and Tulane."

"By 1969, Nixon announced the upper class' party line. He said 'unwanted children in poor families' caused our only problem. The Commission on Population Growth and the American Future was an example of a forum for elites to establish national policy. In 1972, the Commission presented the idea that 'only wanted children should be born.' They did not mention the billions of people already born and using up the planet." Moses' voice began to rise and he stood straighter.

"These efforts by America's elitists kept society's attention on the quality of life of the world's people and away from the serious inequalities in the use of natural resources by the rich and poor."

He turned to face Marilee Birnham. "So, now I ask you, Ms. Birnham, if you have ever heard of any serious demands for population reduction sent through the media to public agencies such as the US Congress, to the United Nations, to the World Health Organization, to the US Department of Health, to the International Planned Parenthood Foundation? To the Association for International Development? Or anyone else? Any call to action about the problem of too many people? Has the general public ever heard that overpopulation causes environmental damage which in turn causes food shortages, ethnic competition, and refugee emigration?" He leaned forward.

"The Population Council has sat and watched overpopulation threaten to destroy our world. My personal interpretation, Ms. Birnham, is that this organization preempted the study of population problems, understated how excessive population damages the environment, and by doing so served the long-term multinational corporations' goal of growth. Growth at any cost, in any country; growth regardless of damage to that country's economy or culture so that Western corporations could endlessly expand markets for their goods and profits."

"Your 'Captains of Industry' are not the solution, Ms. Birnham, they are the problem." Spent, Moses fell back in his chair.

Marilee Birnham rose with her eyes gleaming. "An excellent review that I needed to hear earlier, Moses. I always wondered where you stood. So how do we solve our current problems?"

Jonathan Remmer jumped to his feet. "Marilee, exactly what we hoped has happened. If the hazards weren't so great, we should celebrate. I urge a vote of confidence for our officers. They should continue to seek a reproduction limiting agent through the Birnham Prize."

"I agree," Marilee Birnham said, "and call for the question."

Astonished, Moses Tyler managed to ask, "All in favor?" All hands went up around the table. "The ayes have it." Tyler sank into his chair and put his face in his hands.

Marilee Birnham went to Moses and kissed him on the back of the neck. "You are a great Chair for this Foundation, Moses. Please tell me you will keep up the good work. You have my complete support."

Moses sat up with tears running down his cheeks, and nodded. He thought he had thrown away the chairmanship and his life's work.

As the group headed for the exits, Tim Gedland waved at Jonathan to complain. Remmer waved him away. "Call me when you've read *Beyond Oil*."

16

PARTY'S OVER

Colleen and Jonah soon bonded into a tight work group with their new work mates at Biosolz, Susan and Ted. The hazards and technical problem solving drew them together. Although their associates sometimes called them the "Horsemen," as in the Four Horsemen of the Apocalypse, Mary called them "The Mob," referring to their inseparability and stubborn self-defense.

Sean called the Mob together. The terms of their contract required that they isolate themselves from the outside world until they had production well underway. Now, Mary wanted to know how much information had leaked in. She gave Sean the job of prodding them to find out what they knew.

"Mary called from Europe this morning, and asked me to gather some information. She said that three companies have recently asked us to sell them small quantities of our new product."

Jonah asked, "What companies?"

Sean hesitated. "Bauman Labs, Sievert Filters, and Milbar Biologicals."

Jonah said, "Those are big outfits."

"Mary also asked if any of you could explain how these commercial companies found out about our product? Also, do you recommend that we sell them the virus starter samples?"

"Mary thinks we leaked information to those companies?" Colleen asked.

"I don't know," Sean said. "Anyway, Jonah knew the names."

"Everyone knows those names," Colleen said.

Sean asked, "Anyone have any ideas?"

Susan Trey balked. "I'm offended. How the hell should I know about an information leak? All we do in here is put our heads down and work day and night. When and how could any of us have contacted outside companies? She's the CEO. She knows our contracts forbid that."

"She just asked about the possibility," Sean said. "What do you want to tell her about selling the starter?"

"How much money will they pay for the samples?" Jonah asked.

"I don't know," Sean said.

Jonah pressed. "Take a guess."

"I'd think each sample might each be worth a few million," Sean said.

Shocked, the Mob asked, "Several million? That much?"

Sean said, "So do I tell her yes or no?"

"Tell her no!" Susan said.

"Hold on," Jonah said. "Assuming a total of $10 million, Colleen would get three million, I'd get one million, and you two would each get $200,000. Now how do we feel?"

"Three million?" Colleen asked.

Ted said, "Tell Mary we're thinking about it."

Colleen agreed. "Do that."

The next day, Colleen shut down the virus lab. The four of them spent the day locked in Colleen's quarters, explaining to each other how mistreated they were. What changes had occurred? Why would Biosolz sell starters and take away their monopoly? Why did Mary imply that they had leaked information to competitors? Didn't she trust them? Did she want to ease them out? Why did Mary ask Sean to raise these questions? Why didn't she do it herself? Had Mary sold out? Had something gone wrong? They speculated and griped all day. The meeting became an informal strike.

Biosolz occupied a 130 acre industrial park, about a quarter mile on each side. A double row of tall windbreak trees marked the external boundary of the compound.

Back from the trees, a seven foot hurricane fence stood, monitored by vibration sensors. Numerous spotlights flooded the area at night, and watchmen patrolled the grounds. Security guards restricted access to the main gate. Mary's administrators knew of a carefully disguised exit to be used in emergencies.

The scientific activities were restricted to a fenced compound that covered a quarter of the park. These interconnected modular steel buildings carried a maze of air passageways, air conditioning fixtures, evacuation pumps and filters, concrete posts, steel beams, towers, tanks, pipes, wires, tubes, ladders, and lights. Armed guards restricted access at the gate near the clean-up rooms where workers changed their clothes.

The living and recreational quarters nestled in the privacy of groves of trees.

Mary Henly's domain, the Administration Center, rose two stories; all else stood one-story. She asked the designers for stone and wood construction with a modernistic and imposing high-tech motif.

Mary stalked into her home base, the Biosolz business offices. Her manner changed. She became abrupt, demanding, expressive, even imperious.

She questioned her Executive Officer. "Michael, where is the next auction?"

"Wall Street wants to do it. They'll sell for a thirty minute period at the end of a business day. By the way, how much is there?" Michael asked.

"Fifteen tons of solution, but no one is to know that."

"Of course," Michael said.

"Should we accept?" Mary asked.

"They want 10% of the gross receipts. That's a bit high."

Mary asked, "Will they agree to secrecy?"

Michael nodded.

"It could be worse. What else is there?"

"Close to hand in beautiful downtown Durham," Michael said. "Revelation Tobacco will rent us their auction block to sell 'exotic strains of genetically engineered tobacco starter seedlings.'"

"How do we deliver?"

"The DLH Courier Service will accept our 50 pound five gallon containers and crate and deliver them to our purchasers, on a cost plus basis, to any location."

"They'll ship the crates as part of their regular shipments? What if they have a spill?" Mary asked.

"They don't seem worried about that. Oh yes, they want the right to send bidding agents to Durham for their own purchases."

"That sounds great. Do you agree that we should nail it down?"

"I do, and I'll call now," Michael said. "How's production? How's the Mob?"

"Production is down," Mary said. "The Mob is barricaded in the housing area, bitching about the lack of trust we have shown. By now they may be only a bit fretful. They're about ripe to join a union."

Michael laughed.

Downtown Durham never saw so many limousines. The vacant, dirty, run-down, broken-windowed, ramshackle red brick buildings represented the old tobacco industry and stood in sharp contrast to the modern cars which surrounded the site.

Inside, phone banks connected the bidders with the outside world. At least two hundred financial moguls attended, many with a separate entourage.

With two blows of his gavel, the auctioneer started the proceedings.

"Good day, y'all. We have an unusual product to offer, starter seedlings of a very special tobacco plant, guaranteed to increase your yield per acre, and your sale price per pound. The auction will run according to the published rules which you received in your advance brochures." At this point, they took a 30 minute break to arrange payment procedures with the financial agents, if they were to win a bid.

"How much did the auction yield?" Mary asked when Michael returned.

"We sold 3200 gallons of concentrated virus at about four million dollars a drum," Michael said. "The biggest buyers were China, Saudi Arabia, and Royal British Shale. China's purchase has an add-on requirement. We must provide teams to go there and build virus production facilities for their own use."

"That's interesting," Mary said. "China wants the long term ability to use the virus."

"Sounds like it. India, Taiwan, the International Family Planning Association, and the World Health Organization were next. Then there was a group called the New World Order of Germany. Unsavory bunch if you ask me. Paid up front for the whole batch."

"Excellent," Mary said.

"The remnants included the Detroit Foundation, Exon Oil Corporation, and Disposable Diapers Co. The Skinheads of America, and the IRA offered bids but they were low."

Mary laughed. "Now the diaper company is going to start counter measures against the virus?"

Michael laughed. "Oh yes. The Center for Disease Containment bought a drum."

"What? Instead of confiscating it from us?" Mary asked, astonished.

"Well, it's just taxpayer's money. You know the Feds don't like to hold money back from the private sector," Michael said.

Mary's eyes gleamed as she read the results. "How much of the sale price have we received?"

"About a third," Michael said. "The rest is due when we ship."

"So you banked almost $800 million of this at the close of auction?"

"That's right."

"Good Lord. Who could have predicted? Where is the material now?" Mary asked.

"It's being transferred by tank truck to a packing broker in Atlanta who will seal it in stainless steel 5 gallon drums under sterile conditions." Michael smiled. "He demanded $1000 per drum, which is about a quarter of 1% of our sale price. I let him take advantage of us."

"Beautiful, Michael. Now, what do you hear about our sensitive production crew?"

"Not so good. Sean has complained since you left. The Mob doesn't seem to like him. They quit work."

"Really?" she asked with surprise. "Well, it looks like we need some discipline for our headstrong children," Mary said. She continued to examine the sales list. "Tell me a bit more about what the Chinese want overseas."

"They want us to help them set up a production plant in China. They require a team to get them started. When the plant yields the first production, they will pay the last 40% of the total purchase price."

"Hmm. What does our vaunted management staff think of that?" Mary asked.

"They are overwhelmed and would do anything to get the money. They've lost their judgment, in my opinion. I do agree that we should send a team."

"Good Michael. I appreciate your astute analysis." Mary left.

That means she agrees with me, Michael thought. Thank goodness.

Mary let herself into Colleen's quarters where the Mob met.

Ted saw her first. "What's going on, Mary? First, you use our contract to lock us in the plant for 60 days. Then you have the gall to have your boy friend Sean accuse us of leaking information."

Mary stiffened on hearing, "your boy friend Sean."

"I still don't know how the information got out," Mary said. "It didn't hurt to ask if you knew anything about it. Frankly, I don't care if you feel irritated. I also tested you to see how much profit we should take, and how much you would settle for. As it turns out, we can turn a profit from those three companies of about 30 million dollars."

She stared sternly around the group. "Now, do you want to complain about my testing you, or do you want to talk about taking a large profit. Ted, you and Susan will each realize $600,000 in this transaction alone. Now, what do you say?"

"Good heavens, Mary. It's a fortune to all of us."

"Then speak up, you spineless technocrats. Tell me to jump on the offer."

Jonah spoke to the group. "Are we losing our lead for too small a price? Remember, they may be able to displace us once they develop their own process."

"Yes, but it is too much money for me to ignore," Colleen said.

Mary said, "Spoken like a good capitalist. I'll close the sales today." Mary had manipulated the Mob, concealing the last auction and the three bids for virus starter batches. Now, she had to sell to these bidders quickly before they found out they could buy for less, elsewhere. Mary knew she would be criticized for sharp business practice, but she couldn't give up that much money.

Mary's demeanor changed. Smiling enigmatically, she sat down, made herself comfortable, and told them, "I have more news." She beamed at them.

"The four of you can leave here, and go to help one of our clients set up virus production facilities in their country. For this service, Biosolz will pay each of you two million dollars, plus expenses for up to six months. Colleen will be on separate contract."

She paused a few seconds. "If you can't do the work in six months, you must stay on until you have actually produced virus. I don't see this as a problem since you carried out the same project in less than three months here."

"Where do we go?" demanded Jonah.

"Can't tell until you commit. There are secrecy requirements. Why do you ask? In your young life, there are countries to which you would not go?"

Jonah blurted out, "Yes. Nigeria, and Iran."

"Well, of course," Mary said. "But this is not such a state. This is a larger country which very much needs your services." Turning to Colleen, she added, "The entire operation requires your leadership. You'll be Team Director, with bonuses and perquisites."

Mary looked around the room, amused at their disarray. Susan and Ted sat stunned.

"Well, no hurry. Think about it. Let me know your thoughts in the morning. We have 72 hours to let them know." As a management strategy, Mary forced them to make hurried decisions so they wouldn't have time to consolidate resistance to going. She also liked to watch them struggle. She felt that if they fought the problems she gave them, they wouldn't come up with others she didn't anticipate. They actually had weeks to decide.

When Mary left, Colleen asked the others, "How do you feel about working for a country which has asked for your help?"

"Better than working for this corporate icebox. I don't feel appreciated here," Susan complained.

"Exactly my sentiments," Ted said. "I feel like I'm in a rat maze that Mary designed."

"Yeah, and I'd rather feel my work could help someone," Colleen said. "Speaking of that, what is Biosolz doing with all the virus we produced?"

That night, Colleen told Jonah, "Mary just plays with us. She looks pleased whenever we are shaken up. She talks to us as if we were 10 years old."

Jonah agreed. "She just tried to provoke us when she had Sean ask us about the leaks. I wonder why?"

"Well I don't like it, and I don't like her," Colleen said. "I'd also like to know what's happening out in the real world."

Jonah watched her soberly.

"I want to get out of here," she said.

"We might have to go to a place like Calcutta," Jonah warned. "Are you ready for that?"

"Damn right," Colleen said. She hadn't realized how much she resented Mary's manipulations. "Can you handle it?" she asked.

"I have conditions," Jonah said.

"Really?" Colleen looked apprehensive.

"I want you with me every night, no excuses."

"Hah." She threw herself at Jonah and hugged him. "My god, I had no idea how much anger I've stored up."

"Don't worry about me," Jonah said. "I'm still on the honeymoon. If you're with me at night, I don't care where we go."

Colleen kissed him. "I feel better already."

Jonah was glad Colleen asked if he wanted to go. He knew that she could have decided on her own, and they would have followed.

17

PRODUCT SAFETY TESTING

In preparation for the highly profitable auction in Durham, Mary Henly had hired a marketing agency. They provided her with a glossy advertising brochure and two small lists of wealthy individuals, and wealthy groups that involved themselves in violent strife: ethnic; religious; racial; or nationalistic. Henly made sure that all the names on the lists got news of the auction. The list included a certain Prince Khalid who arrived at the auction and bought several kegs of virus.

Prince Khalid resented the sale of his people's petroleum to the US just so Detroit could sell the cars to burn it. In his youth, like so many wealthy mideasterners' children, Prince Khalid attended college in the US. He had trouble learning English, and his first college years gave him trouble.

"Sir," he said to his geology teacher. "You gave to me the grade of D. My country does not accept the grade of D. I must have the B or they will not send money."

"Khalid, you didn't do very well on the final exam," the geology teacher told him, as he reached for a stack of papers.

"Sir," Khalid said, "I must have the grade of B, so please to change the grade. This D will give me great trouble in my home."

"I can't just change it. Do you want to study and take the exam again?" the teacher asked.

Khalid thought this over and said, "I must have the B, sir."

The impasse continued over several days. Finally the exasperated teacher told him, "Khalid, don't ask me to change the grade again, or I will write a letter to the Dean. They might throw you out of the school." That will teach him a lesson, the teacher thought.

Khalid attended other four-year institutions, and found himself especially attracted to those economic analysts who vigorously criticized capitalism.

He learned the sociologists' three criteria to identify persons who exercise power: those who benefit financially, those who set agendas, and those who control decisions. He became a staunch Marxist, not a friend of the United States. Years later, he gained access to large sums of petroleum dollars from the family coffers.

He thought back to his days in Los Angeles, as he purchased a ton of Colleen's spray at the Durham auctions. Then he approached an American pilot flying agricultural spray planes in Southern California and offered him a great deal of money.

The crop sprayer agreed to make passes over Los Angeles until the spray tanks on his three planes were empty, concentrating the aerosol at a time when most people would be in the streets.

To carry out the contract, the pilot, Jonathan Siegle, went looking for the rest of his team.

He found his two pilots completing a long day of beer-drinking in an Oxnard bar, practicing their Spanish on the waitress. He ordered another round with tequila to loosen them up for the job. "Charlie, who is this diving out of the sun, all guns blazing?" Pete asked.

"It's our glorious leader," Charlie said.

"What does it mean when our fearless friend buys us drinks?" Pete asked.

"It usually means he's promised to do a job so dangerous he's afraid to do it alone," Charlie said.

"Thanks for the drinks, Jonathan. Do we have to spray those Canadian spruce beetles again?" Pete asked.

"It's safe to tell us, Jonathan. What are we spraying?"

"Fellas, it's a new pesticide for the medfly," Siegle said. "But the Feds want us to test it before they license it."

"Down in Orange County?" Pete asked.

"Naw," Siegle said. "The manufacturer wants to make some passes right here in Los Angeles, but quietly. You remember that godawful medfly spray controversy?"

Charlie squirmed. "Why are we going to spray downtown LA for a fruit fly?"

"Actually," Siegle said, "it's a mixture with a new pertussis vaccine for whooping cough. They want to check out both."

"You're lying, Jonathan," Pete said. "Your nose twitches when you lie."

"Well, maybe it isn't worth the money," Siegle said.

"Money? What isn't worth it?" Pete snapped.

"If we make this one flight . . ."

"Yeah, then what?" Charlie asked.

"Then they will pay each of us $40,000 if we spray where I

show you," Siegle said. Although he got $300,000 for the job, he did all the negotiating, he reasoned.

Pete leaped to his feet. "I'll follow you anywhere. We can't let the California citrus growers down."

"I'll lay it all out for you," Siegle said. He left before they could change their minds.

On the chosen day, they flew three planes to an isolated landing strip, just over the hills north of Malibu. Inside a dilapidated agricultural shed, they painted out the aircraft identification numbers and replaced them with those of a competitor's planes in Orange county.

The next morning, the three pilots took off a few minutes apart to cause less attention. They dropped down through a canyon in the hills above Malibu and flew along the beach.

Siegle flew low and started his spray at Venice. Everyone waved as he went over Muscle Beach. Then he turned inland. Charlie sprayed Santa Monica, then followed the freeway East. Pete sprayed Malibu, then East over Hollywood. Two of the three planes went over Prince Khalid's alma mater, the Los Angeles college which gave him a "D" in Geology.

As each pilot returned to the hangar, he painted his plane another color, and replaced the original identification numbers. Then they flew home to Oxnard.

Back at the Biosolz compound, Colleen opened a letter from Regis.

"He wants us to help him with his speech for the Western Division of the National Association of Riflemen."

"Let's hear it," Jonah said. Colleen read aloud.

"We in the NAR should support environmental organizations that hold objectives similar to our own. Though the media labels us as extremists, we only advocate the strict application of laws to assure the protection of animal habitats."

"We strongly oppose the on-going destruction of wetlands, forests, grazing lands, deserts, water resources, and other wildlife habitats brought about by development, municipal use, and farming."

Jonah groaned.

"Thus, we ask that the agencies responsible for protection of the outdoors be funded adequately in order to enforce existing laws. We seek balanced families of plants and animals in nature and oppose their destruction for any reason."

"As your Division Chair, I will seek cooperation with the Wildness Club in recognition of their ability to raise a 40 million dollar annual budget."

"He's losing his judgment," Colleen said.

"They may kill him," Jonah said. "Boy does he have guts."

18

A LONGAGE OF PEOPLE

It was an ordinary hot and smog-filled day in Mexico City, two months after Biosolz auctioned off 15 tons of virus solution in Durham, N.C. Most governments and a few citizens now knew of the Jameson virus.

On learning of the virus, top Mexican Government officials secretly set plans in motion with the help of a Mexico City pediatrician, Juan Fernandez, MD.

Dr. Fernandez directed the Pan American Medical Society of Central and South America. He gathered a handful of southern hemisphere medical specialists for an emergency meeting in Mexico City.

Dr. Fernandez watched the assembled physicians finish the refreshments and the mandatory social period. He had raised the money to pay their way to Mexico City. The visitors wondered where he got it and whether they could get some.

"My friends, could we assemble?" He waved the seven men and two women, all medical doctors, into the hotel's opulent conference room. They required a few more minutes before they stopped talking.

Dr. Fernandez tapped lightly on the conference table. "Could we begin? The program calls for a field trip to leave in a few minutes. May I set the stage for that experience?" God help us, he thought.

"I wish to read some comments from Dr. John Glass of UNICEF, regarding the plight of the world's children." Fernandez adjusted his eyeglasses.

"Each day, forty thousand children die, and a comparable number are disabled for life, the vast majority from causes for which we have long since discovered low-cost cures or preventions."

"Their families are too poor to avoid the malnutrition or disease which kills them. They live in third world countries which often pay a quarter of their export earnings toward interest for debts to the industrialized world, money which is taken from health and welfare budgets." Fernandez looked up in time to see Dr. Rodriguez of Venezuela raise his hand to break in. He appeared to be the elder statesman of the group, perhaps 70 years old, with white hair, courtly manners, and now, an angry glare.

"My dear Dr. Fernandez, the poor have been with us from time immemorial, and shall continue to be. There is a shortage of the world's resources. Surely you accept that?"

"I accept only that the population is too large for our resources," Fernandez said. "Furthermore, I wish to propose a remedy, an experimental program, to be applied to Mexico City, where I myself practice and teach. How many have heard of the new 'Jameson strain,' a virus recently discovered by a US virologist, which blocks spermatogenesis?"

One member asked, "Why? Is it spreading?"

Dr. Enrique Anaya, from Chile, said, "I don't think that is why Dr. Fernandez asked. You contemplate inoculation?" Anaya had darting, probing eyes and the quickest mind. He was a young neurologist, a researcher, and an ideological firebrand, still furious over the American CIA coup that killed his hero, President Allende.

"Gentlemen, before pursuing the topic, could we proceed on our field trip?" Fernandez asked. "We can continue the discussion on our return, in about ninety minutes? Thank you very much." He moved the group out to a waiting bus.

NO POT TO PISS IN

The tour bus wended its way through thick traffic, heading for the Zócalo, Mexico City's great central square. Ms. Teresa Quintana, a social worker and colleague of Fernandez, came along as tour guide.

Fernandez spoke quietly to Dr. Emilia Victoria. "Maybe my critics are right. Am I too impatient with these problems?"

"Juan, keep pushing. Most of us burn out," she told Fernandez.

Fernandez chose pediatrics as his sub-specialty because he knew how much children could be helped if you started early in their life. He deeply resented society's treatment of children. For him, poverty was child abuse.

The bus left the square of Zócalo and headed for the nearby

barrio of Morelos. They jolted over dusty, pot-holed streets past continuous small gloomy entrances leading to a rabbit-warren complex of patios.

Ms. Quintana directed the driver to stop at an entrance point. Dr. Fernandez got out. "This is my colleague, Teresa Quintana, who will take us through the quarters," he said to his guests.

Ms. Quintana took them through a patio or open space. The smell of sewage and grease filled the air. A jumble of ten rooms extended from each side of the patio. Each room held a half naked family, perhaps five people. Babies cried and small children peered out of doorways, most with no clothing or shoes. Flies buzzed continually. Quintana took them toward the back where two overflowing bathrooms served the hundred people in that section. Two of the men hung back, retching.

Ms. Quintana finally led them back out to the bus. She spoke as they drove along the outskirts of Mexico City on a tour of the shantytowns which housed tens of thousands of people.

"The people in the Morelos barrio that we just left are considered lucky," Quintana said, pointing to the new scene.

Quintana described the huts made of cardboard with tin roofs. "They have no running water here and no toilet facilities. The stench comes from garbage and human waste lying nearby in growing piles." Quintana bounced along imperturbably.

"Rats roam freely like stray domestic animals." She turned to Fernandez. "Do you wish to stop?"

Fernandez shook his head. He couldn't bear to get any closer to the two children he saw fighting over some unrecognizable scrap. Other children of all sizes picked through the piles of refuse, occasionally placing something in their mouths. Small children wandered about as far as one could see.

Ms. Quintana turned from the scene outside. "Mexico City has more than twenty million people, doubled from twenty years ago, perhaps the most populous urban center on earth."

As they returned to their hotel, Ms. Quintana took her last opportunity to give perspective to the city. "Mexico City has been described as an ecological Hiroshima, struck by a population bomb." Bless you, Quintana, Fernandez thought. She spoke with the deep outrage of a dedicated public health worker.

As they debarked from the bus, Fernandez spoke to them. "Gentlemen, I claim that we do not have a shortage of resources. I say we have a 'longage' of people. It is noon, now. Could we rejoin at 4 p.m., back in the conference room?"

Without a word, they withdrew to their separate rooms.

As Fernandez headed back to his room, he saw his antagonist, Dr. Rodriguez, at the bar tossing back a stiff drink. Fernandez decided to confront him. He sat down beside the older man and ordered a drink. "Sir, as I recall, you are from Caracas?"

Dr. Rodriguez turned and glared at him. "That is correct." He frowned, shook his head, and ignored Fernandez for a while.

Finally, he spoke. "I don't like what you did today. You remind me of the insufferable young communist pups I have known."

"I'm really apolitical," Fernandez said.

"However, your tour disturbed me. You have my attention. I look forward to hearing what you think we should do, as physicians."

"Thank you. I appreciate that," Fernandez said.

Rodriguez finished his drink. "I shall retire now. I'll see you in the room at four." He nodded formally as he left.

AN ECOLOGICAL HIROSHIMA, STRUCK BY A POPULATION BOMB?

Back in the meeting room Dr. Fernandez noted that all nine physicians returned for the battle.

He asked, "Does anyone object if we do not take minutes? I would like all of us to agree to keep these discussions confidential."

No one objected.

"I have asked Dr. Rosa Carrera, our Associate Director, to help with this presentation. Dr. Carrera, would you tell the group how our meetings started?"

"Thank you," she said. "Two months ago, the Mexican Government asked Dr. Fernandez and me to study a major public health issue. We agreed and attended a three-day conference, arranged by the President of Mexico."

"We were properly impressed," Fernandez said.

Dr. Carrera continued. "The Mexican Department of Health wants you to review a plan. You represent many medical associations, and the government wants your opinions. You don't have to approve the plan. But they would like you to examine it and say that it makes sense and the objectives are humane." She turned to Dr. Fernandez.

"These health officers knew about the Jameson virus recently discovered in California, which blocks male fertility." Fernandez paused. "They propose to spray the virus over major portions of Mexico City."

The group sat stunned.

The psychiatrist Pedro Herrera finally said, "You can't be serious."

"Entirely serious, I assure you," Carrera said.

"But to attack the males of a society and destroy their masculinity?" Herrera said.

"I wasn't aware of hormonal effects. We're talking about blocking their breeding ability," Carrera said. "We are not trying to prevent intercourse."

"My God. What have these men done that you would attack them this way?" Herrera asked.

Dr. Carrera said, "Well, let's look at this group of males. What are the facts?"

Dr. Fernandez spoke. "I am most concerned over the reports of medical anthropologists that poverty stricken young Mexican males past 13 or 14 don't get feelings of self-worth until they have fathered a child. These boys' own parents can't support them. They may try to live with relatives. They can't get jobs. They have no possessions. Yet they don't really consider themselves men until they've gotten some 12 or 13 year-old girl pregnant."

"They often father a child by the time they are 14," Carrera said.

"Let me understand this," Dr. Herrera said. "The only feelings of worth for these young men come from starting a family?"

"No," Carrera said. "They don't have the needed resources to start a family. All they can do is get a girl pregnant."

"But you are planning to prevent this?" Herrera shook his head. "Speaking as a psychiatrist, I promise a strong adverse reaction from these men."

"Boys," Carrera said. "Fourteen year-old boys. They can't get jobs. They can't feed themselves. They certainly can't feed a mother and baby. They are barely into puberty, and they are emotionally immature as far as caring for a child is concerned."

"But, to sterilize them?" Herrera said, "It is vicious."

"How vicious?" Fernandez asked. "With all these babies having babies, almost half of the city's 20 million people are age fourteen or younger."

"Dr. Herrera, let me ask you something," Carrera said. "If the virus blocked conception in females, would you feel better?"

"Well, you didn't claim that women get their self-confidence from having a baby. If so, sterilizing them would be less damaging socially."

Now Emilia Victoria spoke up. "So it's alright to sterilize women but not men?"

Herrera refused to answer.

"This sounds like ordinary male chauvinism to me," Dr. Victoria said.

"Let me give you more data," Dr. Carrera said. "The situation gets worse. These fourteen-year-old boys are trying to get a girl pregnant so they can say they are men. But if she doesn't get pregnant, they say it is her fault and blame her. They may abuse her and leave her feeling useless and at fault. Then they move on to the next victim. Whatever happens, these girls lose the most."

Fernandez tried to redirect the discussion. "If we claim we care about all these babies born to these children, then we must note that since 1990, the infant mortality rate has gone up. Hundreds of thousands of children are abandoned each year in Mexico City."

Dr. Rodriguez asked, "What causes the mortality increase?"

"As we said, the young mothers are undernourished, and there is no health care available for many," Emilia Victoria said, still outraged at Herrera.

"The smog is a major cause of poor health." Carrera moved to a window and pointed out to the thick haze which obscured the horizon. The haze particles turned the incoming light into all shades of red.

Fernandez explained. "Mexico City is high, 7300 feet, and suffers from an inversion layer that traps lead, smog, and other toxic materials plus a good quantity of powdered feces in the air breathed every day. Even the city's lenient standards for ozone pollution are exceeded over 8 days out of 10. Mexico City has the filthiest air of any city in the world. Foreign female diplomats who become pregnant are told to go home while they have their child."

"We like to tell people that early child care is important," Carrera said. "But seventy percent of these babies will be born with levels of lead higher than World Health Organization recommendations."

"What does that do to the children?" Rodriguez asked.

At least the old man has enough courage to ask, Fernandez thought .

"It may mean that a whole generation of Mexican children may be intellectually stunted," Carrera said.

Several heads shook in dismay.

Carrera continued. "Air pollution will not get better. Many U.S. corporations have moved to Mexico because of the lax air pollution laws. The NAFTA agreements of 1994 increased annual toxic emissions into the air by 11 million pounds in Mexico and even more in the US."

"There are other problems," Fernandez said. "Five million

people each year could be fed with the food lost due to the annual loss of soil in Mexico. That's a quarter of the people in Mexico City."

Carrera concluded. "We should add that women devote 100% of their income to buy food for the family. But adult male household members may spend a quarter of their income on wrist watches, radios, bicycles, things which take food away from the children."

Carrera and Fernandez stopped and waited for a response.

Finally, Dr. Rodriguez shook his white head. "I find this appalling. You make me wonder what my own country is tolerating in Venezuela."

Rosa Carrera, from Sao Paulo, Brazil, said, "At home, our women want access to contraceptives. Mexico is similar, and I believe many would welcome male sterilization."

"There may be a more fundamental question," Carrera said. "Do Mexican government officials really worry about poor people? In Brazil, our politicians don't seem to give a damn about them."

Most agreed that there is a lack of concern.

"Your politicians may only be worried about social unrest, uprisings, and revolutions," Anaya said.

Trust a Chilean to understand that, Fernandez thought. "I think you are right. Elected officials have already admitted that no one can control Mexico City. They also don't want to spend money on the military. Mexico spends less than 1% of GNP on her military forces, while Nicaragua, for example, spends over 10%."

Anaya continued. "Then I'll bet they're running scared. Both Brazil and Mexico have a large upper class which is 20 times as wealthy as the bottom 20%. The political scientists warn this is unstable."

"I think you are right," Dr. Victoria said. "The Chiapas uprisings frightened Mexican government leaders."

Fernandez turned to Dr. Rodriguez. "Sir, this morning you challenged me to provide a rationale from the point of view of physicians for holding down population growth. Do you believe it is in the public interest to begin this project?"

The old man slowly stood up, his brow furrowed with concern. He polished his glasses and arranged his lapels, cleared his throat, and appeared to make a decision. He spoke slowly and formally.

"I was not convinced this morning. However, the situation you have described is quite unacceptable. No responsible professional could ignore these disastrous trends. I guess I am saying that I support this effort, at least within this group, if not publicly."

"My only concern," Emilia Victoria said, "is that the project keep a low profile."

"Definitely," Fernandez said. "The administrators in the Mexican government do not want an irate populace coming after them."

"I still fear massive unrest when the effects are known," Herrera said. "However, I have not heard a more distressing justification for what you propose. I never thought I would hear myself agree to this."

Dr. Fernandez then told them, "The President's representatives asked that I serve as their agent and set up the necessary contracts for the project. They wanted you to review the plan to assure that it is worth doing."

"Where will you get the money?" Dr. Anaya asked.

Carrera said, "The Mexican oil companies will provide the needed funds. They prefer controlled population limitation to continued growth with social unrest. The uprisings in southern Mexico worry them."

Dr. Fernandez added, "As proof of good faith, they provided me with a checking account and a start-up sum and a telephone number when more is needed. Dr. Carrera and others will monitor our progress."

They discussed the need for a reliable overseas commercial contractor, whose country's laws give anonymity to their clients.

With this consensus, Juan Fernandez, MD, agreed to start the project. He began the search for project designers from his temporary base in Southern California.

19

HEAR NO EVIL

The Deputy Director of the FBI met with the US President in the Oval Office after they completed the investigation of the Los Angeles aerial spray.

The President immediately demanded to know, "Who developed the spray?"

"A Berkeley doctoral candidate named Colleen Jameson. She was top of her class."

"Why in God's name was she trying to develop a virus to sterilize people?" the President asked.

"Sir, she claims she just happened upon it."

"Did she have any company affiliations? Corporate scholarships?" the President asked.

"Not that we could find. Sir, we located the pilots who sprayed Los Angeles. The Department of Health worked them over, since the pilots ran live virus solutions through their agricultural planes."

"Who paid them?"

"Someone paunchy, about five feet ten inches with a swarthy complexion, dark hair, a gold tooth, and a bushy mustache. He spoke English maybe with a Middle-East accent. He placed an ad, and told the pilots' business agent he had a fruit-fly spray job. He gave them the barrels of spray, paid them $150,000 and disappeared. He sent them the balance of the payment afterward."

"Did you find all that plausible?" the President asked.

"We thought not, at first. The pilots were paid a lot of money, and we tried to come down hard on them. But they still couldn't give us the name of the man who commissioned them. I don't think they ever tried to find out his name."

"What the hell? Has everyone given up their scruples?" the President asked.

"It appears so. The pilots were angry that we bothered them. They didn't see this as any different from what they do all the time. They told us bitterly that the agriculture corporations which give them contracts often pay them to do illegal sprays. Wrong pesticides, wrong dilutions, wrong surfactants, wrong wind conditions, violations of EPA standards, safety violations, violations of flight rules, you name it. They were ready to give us names, dates, and job orders to document those violations, but we let it go. We had enough to worry about."

The President shook his head in dismay. "You must be right. The media and the residents of Los Angeles didn't complain much. Maybe they expect that sort of thing in LA."

Two weeks later, the President got a visit from the Mexican Consul General who asked for a private non-talk.

When the Consul General left, the President called in his National Security Adviser, and charged him with secrecy.

"The Mexicans are spraying their own people with the Jameson virus from California."

The President's Security Adviser asked for time to make inquiries to help decide what the President should do. Two days later, he returned.

"Mr. President, the Business Lobby asks that you visit with them next week. As a relative newcomer, you haven't had the pleasure of their advice. You'd better go."

"That sounds like a threat. What do you know about them."

"They formed 20 years ago. They spent a lot of time in the White House during Reagan's reign. The group consists of the Chief Executive Officers of the largest 200 corporations in the US, and members may not send a substitute to their meetings. These are the most powerful people in the US. They have strong-armed every president in office since they formed."

The President met with members of the Business Lobby, then called a few key staff members together for a briefing. They had never seen the President so dismayed. He told them, "I must share with you some orders I've been given." An ominous beginning, an aide decided.

"I met with six corporation directors who said they spoke for the Business Lobby. Above all other considerations, they don't want us to intervene in any foreign countries, including Mexico, unless it benefits their business arrangements."

"Their basic goal is to arrange for a global factory, with no restrictions on what can be sent anywhere, especially where toxic materials are involved. They mentioned a lot of examples. They don't want any regulations involving car emission standards, fishing practices, or produce diseases; also, no specification of levels of pesticides in food, or levels of antibiotics or hormones in meat. They don't want any regulations involving energy, timber, minerals, or water sources."

The Secretary of Commerce said, "Those are all the business agreements from the GATT Uruguay talks. These guys are against subsidies for renewable resources or the efficient use of resources. It's as if we're supposed to use up the world's resources, the faster the better."

"I think several of the corporation directors who talked to me could be contractors for the Mexico City spray project," the President said.

"My God," an aide said. "So why not just ignore these people?"

"I called the National Chair of the Democratic Party and told him about our discussions," said the President.

"He pointed out that each of the 200 corporation CEOs of the Business Lobby sits on the board of directors of a half-dozen or more large corporate and financial institutions, law and insurance firms, media corporations, foundations, and universities. These CEOs thereby directly influence perhaps a thousand organizations." The President looked up and raised his eyebrows for emphasis.

"Here, I'll read what he told me. If I intervene, 'not only will Democrats find themselves without campaign funds for the next election, but the newspapers and television will suddenly discover that Democrats don't cooperate with business, that we refuse to stimulate development or create jobs. Economically speaking, everything that goes wrong in the US will be my fault. My ratings will plummet, the media will make sure of that.'" The President threw down his notes.

"Nixon told us this kind of bullying went on up in the California redwood country."

"So what do we do about Mexico City?" asked the Secretary of State.

The President sat quietly for a moment with his head down. Finally he looked up and said, "We will be deeply outraged at the spraying of Mexico City."

"That's it?" the Secretary of Health asked.

"No, you can start composing some news releases about mistreatment of the underprivileged poor. People will complain, but nobody really wants us to do anything."

"Dammit, sir." The Secretary of Health stood up and glared at his boss. "We ought to do more."

The President returned his gaze mildly. "Phil, have you ever been to Mexico City? Have you seen how bad it is?"

"No," he growled.

"Well, go tour the place, unofficially, right away."

"You are asking me to go to Mexico City?" the Secretary asked.

"No. I'm ordering you to go within a few days. Take your wife and talk to her about what you see. Make our Consul take you around. Smell the air. Check the slums," the President said. "OK, Phil?"

"No it's not. I don't like being ordered around."

"You'll need information if you are going to fight me. Take the trip, and come see me as soon as you get back." The President waved his hand to dismiss them.

"One more thing," he said as they left. "No news leaks. None."

20

HEIR SPRAY

Mary Henly summoned Colleen in great excitement. "We have a hurry-hurry job request. Do you know a pediatrician named Juan Fernandez?"

Colleen shook her head, No.

"He wants you to represent Biosolz during a short meeting in California," Mary said. "OK?"

"What kind of meeting?" Colleen tried to conceal her pleasure at getting away from Mary and escaping the Biosolz compound for a bit.

"Dr. Fernandez wants you to provide technical information to his planning group. They are drawing up a very large purchase order for us." Mary looked excited.

Great. Colleen smiled to herself. Now Mary wants something. "When do I leave?"

"I have your tickets to John Wayne airport. You'll be in Southern California for two nights, then back here."

Dr. Juan Fernandez arrived at the infamous Los Angeles International Airport, and enjoyed the Los Angeles smog and freeway crush. By comparison to Mexico City, LA was clean and orderly. After a short drive, Fernandez found himself escorted into a high priced planning contractor's conference room. He immediately located Colleen and went to thank her for coming.

Colleen watched the planning group staff turn on equipment all around her.

"What's all the gear for?" she asked.

One of the operators said, "The video camera gets a record of conversations and material on the chalk board. I use two computers to build the spreadsheet and the written outline at the same time. It's faster."

The planning director broke in. "Her job is really quality control. If she can't write it, we must try to clarify it. She converts what we talk about into exact procedures."

After introductions, the director presented the project goals:

Objective 1. Within a 12-hour day, to spray 25,000 gallons of a water-based solution over Mexico City, along specified flight paths.

Objective 2. Provide such sprays on a continuous basis, to service any populous area in the world.

Colleen's body stiffened as she sat up in the chair. "What is in the 25,000 gallons of water spray?"

Dr. Fernandez leaned toward Colleen. "Your virus, of course."

"My virus? To be sprayed over Mexico City?" Colleen's voice ended on a rising note.

Fernandez watched her closely. "Yes," he said.

"Is this a problem?" the planning director asked.

"The spray is meant to drift down on the people who live there?" Colleen asked. "How many are there in the city?"

"Perhaps 20 to 25 million." Fernandez said.

Colleen looked in dismay at the circle of faces around her. "Surely you can't bring this about?"

"We will certainly try," Fernandez said. "That is the purpose of this meeting."

"You are going to sterilize the male citizens of a whole city without their permission?" Colleen asked.

"Without their permission or their knowledge. We hope to conceal our role in the operation," Dr. Fernandez said. All eyes remained on Colleen.

"But you can't," she said. "Surely someone will stop you?"

"Who?" Fernandez asked.

Colleen thought first of international authorities. "The United Nations World Health Organization," she said. Fernandez immediately answered:

"The World Health Organization's top medical administrator gave me his discreet approval for the plan yesterday."

"Discreet approval?" The meaning of the words seemed to elude Colleen. "Mexico will attack you as immoral racists and fascists."

"That's right. The President of Mexico will complain the most loudly," Dr. Fernandez said. "And he is the one who hired me to carry out this plan."

"Then I'll stop you," Colleen said, her face pale. She gathered her notes to stand up. "I'll go to the newspapers. The FBI. Someone."

"Uh, Ms. Jameson, may I say a few things," the planning director said. "First, let me say that I have the highest regard for your work in developing your virus."

"It is not my virus!" Colleen's face reddened and her voice rose to a shout. "You think I created a virus like some techie trying to design population control devices? Well, I did not. I stumbled upon it, quite by accident, as it occurred in nature."

The planning director shrugged. "Nevertheless, your name is attached to it. And since you were instrumental in producing large volumes of the virus at Biosolz labs, I think you will be viewed as the designer of this population control project." He punched the top of the desk as he said "this."

"But why should I help?" Colleen asked. "I'm a scientist, not a social worker."

"You are a scientist who has earned a great deal of money up to now," the planning director said. "And this project will radically increase the profits you deliver to Biosolz."

"But I didn't join Biosolz to make money. I was fleeing for my life," Colleen protested.

"Of course you were," the planning director said. "Nevertheless, I think the press will cast you as the leading racist, fascist, and immoral practitioner of eugenics."

Colleen paled and sat frozen.

Dr. Fernandez asked Colleen, "You are aware of the damage done to the Mexico City population by the crowding?"

"Of course," Colleen said. "Biologists often see population blooms. We must recognize them where they occur, even among humans."

"Eutrophication," Fernandez said. "So what is the problem?"

"It's one thing to know about a problem," Colleen said, "and another to spray the whole population from the air."

One of two women in the planning group spoke to Colleen. "Ms. Jameson, you may not know that this plan includes several alternatives. Could we show you some?" The two women brought their paperwork and sat on each side of Colleen.

"Here are two strategies we rather like." Colleen read the two objectives:

1. Sell throughout Mexico City a cheap spray bottle labeled "men's cologne" or "breath spray" or "aphrodisiac," all of which carry the virus.

2. Load a fleet of trucks with crates of bottles which clearly

claim aphrodisiac qualities for the spray. Abandon the trucks in the city where they will obstruct traffic. Allow the locals to loot and sell the "aphrodisiac" laced with virus.

Colleen looked a bit more calm.

"But these young Mexican studs value their machismo life style."

"They certainly do." The two women planners laughed. Colleen looked at them and finally smiled. None of the women liked the machismo culture of the Mexican male.

The planning director spoke. "Ms. Jameson, we really do need high-level coordination so we can deliver the Biosolz virus in viable form. I hope you will help us?"

On the plane back to Raleigh-Durham, Colleen vacillated between apprehension and hope that the project really might do some good. Anyway, I'm not going to help these crazy people, Colleen promised herself. Jonah's going to have to shepherd this one.

21

SPRAY, INC.?

Next, Jonah flew to Southern California to attend the Mexico City Spray Project meetings as Colleen's replacement. He wondered why she complained so bitterly over the project. Didn't she know that most weed and insect spray programs poisoned people? They existed because they made money for spray manufacturers.

In Los Angeles, Dr. Fernandez and the planning director talked all morning to Jonah. They told him to keep the delivery of live virus as the first priority during the assembly of the corporate task force. Then they sent him off to Houston to meet the vice-president of a petroleum equipment corporation.

Jonah found the downtown Houston building and worked his way through the morning crowd. The designer built the expensive and garish entry to impress viewers with the amount of oil wealth manipulated inside. Next came the chromium plated escalators and elevators to the upper floors. Jonah jammed himself in among the highly polished male clones of the standard MBA model, bright eyed, and aggressive, talking in corporate-speak, quick to demonstrate superiority in all matters.

The women among them conformed to a carefully groomed tall Texas beauty standard, and modeled immaculate tailored suits. One such receptionist asked Jonah to wait; she was perfectly posed and poised and spoke near-standard English. She walked Jonah into her boss' office, a white room with black padded chromed chairs and desk, a green blackboard, and a floor to ceiling window that vibrated and threatened to blow out in the frequent winter storms. She immediately poured coffee.

"John asked me to make you comfortable while you wait, Mr. Jonas."

When John arrived they talked about the project without dwelling on the nature of the spray. Jonah focused on a requirement which Dr. Fernandez had emphasized repeatedly. "They want to start up a project away from the US. No visible connections."

"What are the requirements?" John asked.

"They need a home base where they can register and operate boats, small planes, trucks, paving equipment, fuel and water tanks," Jonah said. "They want the country to take our money, register and license us, shield us from inquiry, and ask no questions."

John recommended registration in Venezuela, which had its own oil industry. They could pose as a supply subsidiary. "I think you should get a big oil company specialty corporation to set things up."

"Why a specialty corporation? Why not the oil corporation itself?"

"Too narrow a focus. The big oil companies have pumped other peoples' oil for a century. They have well-paid relationships with politicians, the military, and other businesses in these countries. Some of the biggest multinational corporations in the world are oil companies. British, Dutch, and a few American expatriates often run these operations."

"What's an expatriate?"

"Someone living outside of his home country. You'll sometimes see one who has been gone so long he forgot who he was and lost any scruples he ever had. Maybe he went native, moved in with the local women, ate the local foods, and got the local diseases. Many seem like lost souls: men who drink in bars every night; Americans who take their vacation in Holland because drugs are cheap; Dutch who prefer Portugal because of the weather or the wine. The oil companies are full of dislocated people."

"So who do we approach?" Jonah asked.

"We go to one of the big petroleum corporation service companies. Royal British Shale, for example, has a Houston subsidiary called Pectin, Co."

"Why go to Pectin if the parent company has the money?"

"They're easy to mobilize. They can sub-contract. It's a holding tank for cast-offs who are overpaid and under worked: out to pasture."

"Cast-offs?"

"Yeah," John said. "If you are working in the Nigerian swamps and offend a local administrator, you might be transferred to, say, Brunei. If you screw up there, they send you to Pectin on a make-work project. You'd find yourself surrounded by people who got in trouble over health, drinking, sex, money, attitude, emotional problems. What did I forget?"

"But these big outfits have all the money in the world. They're not super efficient? Training? Counseling? Rehabilitation?"

"C'mon Jonah. Ninety percent of the problems in the American business work force are caused by inept managers. Most of these top administrators are money-blinded incompetents." He snorted.

"Look at my own boss," John said. "He tells me the company is broke, I can't have a raise. Then he takes me to an expensive dinner and brags about the $90,000 after-taxes Mercedes he just bought his girl friend." He shook his head. "He thinks that since she's willing to screw him and tell him how smart he is, he must know how to run things. In fact, he's a piss-poor manager."

"One last question," Jonah said. "If I don't like how things are going, can I redirect these people? Will they listen to me?"

"Sure, just cut off the money. Then they listen good."

"But I can't do that," Jonah said. "Can't I appeal to their principles? Their better judgment?"

"I've never seen it happen."

The Pectin Company expatriates welcomed John and Jonah. Jonah decided they had no loyalties to anyone. They only wanted the money.

Jonah and John needed a few days to show Pectin what they wanted. The proposed mission excited the director, a Britt named Flavin Taylor. He hadn't done anything useful since the company pulled him off a North Sea drill platform after a "cock-up," what an American would call a screw-up.

Flavin Taylor liked Houston, and since Pectin did nothing of consequence, he didn't have to show a profit to keep his job. He set up a training environment where Royal British Shale sent their executives for rest and rehabilitation. If they called it training, they saved taxes.

Taylor arrived early to prepare the Board Room for the visitors. Each of the forty executive chairs carried a small desk and console. They provided all the comforts of first class on a luxury air liner. Like NASA rocket seats, they contained sensors that monitored the physiological responses of the sitter, each chair a high sensitivity interest, lust, or lie detector.

Taylor savored the effect he hoped his words would have. He boldly invited Royal British Shale to attend without prior discussion. What if they didn't show up?

Taylor had just flown in from his winter quarters on Hilton Head. His uncivilized corporate colleagues didn't like the southern plantation ambiencc so he had to fly to Houston. In the spring, cold winds howled among the skyscrapers.

He dressed in his trademark, a flannel lounging jacket and slacks, made slightly severe with the black and cream colors. Women poked fun at his attire. "All ready for bed, are we?"

He had a manicure and hair trim every day. His barber shaved him last. As she worked on him, she mewed quietly in his ear in ways she knew he liked: "Lovely, the profile needs just a bit of . . . ah, yes. Dashing. Be gentle today, my dear. The women will be helpless in your hands." Swashbuckler, he thought when he looked in the mirror.

The others began to arrive. As always, Charlie Drummond, Planning Director for American Central Bank, came in last. "Hi, Flavin. I see you still have these overstuffed Mickey Mouse chairs. Won't Royal Britt spring for some real ones?"

Taylor said, "Probably would, Charlie." Blast the rude son-of-a-bitch. Taylor began the discussions.

"I'm the Chief Executive Officer of Pectin Service Corporation, hopefully for one more quarter," he smirked. "I extend my gratitude, and that of the Board of Directors who authorized this invitation, for your willingness to attend on such short notice." Taylor was not embarrassed that there was no Board of Directors, or that, once a year, Royal British Shale in London called to make sure he was still alive and they still had to pay him.

"On receipt of the proposal, which I will describe in a moment, we decided to ask for the cooperation of our most prominent banking institution." He waved toward the Central Bank group. "We also needed the skills of our colleagues in the Detroit Foundation Management Group."

"What's up Flavin?" Charlie Drummond broke in. "My plane leaves at two."

"Of course, Charlie." Drummond is still the same old bastard, Taylor thought. He launched his pitch.

"We've been approached by some prominent medical leaders from Latin America. They want our help to reduce their population pressures. Some of their doubling rates are down to thirty and forty years. They would like us to attend to some of the festering sores on the body of several countries starting with Mexico City."

"Attend to?" Drummond asked.

"Yes. You probably have heard that there is a new biological agent loose these days. Maybe a virus. At any rate it plays hob with men's testicle plumbing and leaves the recipient unable to reproduce, maybe for life. We've been asked to set up a task group to provide virus spray missions."

"What's the deliverable?" Drummond asked.

"A task group that can spray the anti-reproductive agent over large areas, including much of the Moslem belt from North Africa through India. Also, some populous parts of China," Taylor said.

Warning flags went up for Jonah when he heard the words "Moslem world." Good God, is this how the Christians planned the Crusades centuries ago? He sat frozen faced, fearing the worst, and along it came. "Parts of China?" Now he realized they were talking about more than half of mankind. What next? He waited, aghast.

"Ambitious," the Director of Logistics said. "Any chance of starting a brush-fire epidemic? Work its way out of the third world and really do some damage?"

"I doubt it," Taylor said. "When they sprayed virus over Los Angeles, it drifted sideways before it reached the people below. But infections were limited to that area."

Jonah realized that "doing real damage," meant infecting the industrialized world. Don't worry about the third world. Finally, a Detroit psychologist asked a question that Jonah wanted answered.

"Aren't you worried about public outrage?"

"Well, what can they do? We'll certainly provide enough money to guarantee the cooperation of local authorities. Riots or uprisings might cause problems, but the contractor will be long gone with the money in his pocket." Taylor moved on.

"The long term plan calls for an aerosol spray from airplanes over areas larger than the usual targets of crop sprayers. For distant targets such as Calcutta, Cairo, Lagos, we need a mobile fleet of crop spray planes. Spraying those cities requires a large volume of fuel and spray."

"Preparing the spray will require a technical group with special tanks to provide for dilutions and composition and temperature control." Taylor pulled down a series of maps showing Mexico City about 500 miles south from the southern tip of Texas.

"If you agree, Mexico City will receive the first contract. We'll lease 100 acres of land a few miles east and enlarge a road to serve as a landing strip for agricultural planes."

"We'll bring in fuel tankers and maybe buy fuel locally. Ten to thirty ag spray planes will fly in to the headquarters strip. The crew will be in Portakabins. The virus supplier guarantees spray delivered as needed."

The Detroit Group asked who provided the virus.

Flavin began, "I can't tell you yet . . ."

Drummond broke in. "Flavin, tell them who is providing the biologicals."

"Right." Taylor's jaw clenched. "Biosolz, out of Chicago and North Carolina."

"Are they reliable?" the Detroit Group persisted.

"I'm told they're flaky, but we can watch them," Drummond said. "If we have to, we'll buy 'em out. Right, Flavin?"

"Uh, your information is . . . the same as mine."

"Tell them who is putting up the money for the entire job," Drummond said.

"Before we tackle that . . ."

"OK Flavin, I'll tell 'em," Drummond said. "My informants in the Pan American Medical associations tell me that the Federal Government of Mexico will pay the bills. It's Pimex money."

Taylor fidgeted. "Charley, they didn't want to be identified. In fact, it's essential that the spray operations appear to originate from another country. We'll register in Caracas. We must have secrecy." Taylor peered worriedly at Drummond.

"I won't tell anyone, Flavin. Will you?" Drummond continued. "American Central Bank will support expenses through purchase of the spray. We have to recover 27% per year on the dollar. Royal British will go along with that, won't they?" He turned to the Director of Logistics for Royal British Shale.

"I don't see any problem, Charley."

"Oh yes, Flavin," Drummond said, "control of all project activities will be in the hands of a team under the Detroit Foundation Management Group and our auditors." He turned to the two young staff members representating the Detroit Group.

"OK, you guys?"

"Sure, Charlie. Sounds like fun."

Drummond turned to Taylor. "OK, Flavin?"

"Certainly, Charlie. Whatever you want," Taylor said. "Well, thank you gentlemen for attending. I guess that winds things up. We have food and drink in the next room if you'd like to . . . "

The other members ignored and avoided Taylor, closed their brief cases, and talked with each other as they left the room. A management psychologist from the Detroit Group congratulated Flavin on a well-run meeting. Flavin felt pleased even though he knew management psychologists routinely reinforce everyone.

Drummond stopped with an afterthought. "Hey, Flavin, how about an invitation to Hilton Head? My wife would like to see the place and play golf. Could you arrange that? Stay at your place?"

"I suppose so, Charlie. I'm pretty busy this time of year," Taylor said, trying to refuse.

"That's great, Flavin. I'll give you a call." Charlie pumped

Taylor's hand. As he turned to go, he said, "Have you ever considered going for an MBA degree? I think the University of Houston has a program right here."

Flavin Taylor turned red. "Goddam it Charlie, I have a bloody MBA degree."

Drummond looked surprised, "Is that right? I'll be damned. Well, it was just a thought. I'll call you later about our visit."

Flavin sank into a console chair, now pale and speechless. He sighed. No need for lie detectors to find out what these people thought.

Jonah sat overwhelmed, reviewing what he'd heard. This was the most grasping and unscrupulous bunch he'd ever met. Now, like Colleen, he knew he would have a hard time working with them. Best not get in the way of capitalists when they smell money, he thought. Biosolz is flaky? They'll buy it if they have to? Wait'll Mary Henly and Colleen hear that. He wondered if he dared tell Colleen about the plans to spray the Chinese and the Muslim population of the world. He flew back to Raleigh-Durham.

22

EUNUCHS, ESQ.

Regis Jameson broke away from his legal practice in San Francisco and flew in to see Colleen and Jonah at Biosolz.

"Lots of new problems arising from the virus," Regis told them cheerfully. "We're busier than ever."

"Vultures following ambulances," Colleen snorted.

"Well, people have new virus-related problems, and they ask us for help," Regis said. "You want us to turn our back on them?" He continued.

"Either of you guys want to go with me? I'm off to the American Association of Attorneys' meetings on virus-related problems."

Jonah and Colleen declined, and Regis flew out that night.

The nation's attorneys either wanted to go to meetings in very prestigious locations where their spouses could shop and play, or they wanted them in Washington, DC. The new Ralph Nader University Complex offered both, with commuter helicopters and taxicabs to the Capitol.

In preparing for the AAA Convention, the program planners sent out a call for papers and received mostly new topics.

Clients sought reimbursement for being sprayed, especially when divorced for their sterility. Manufacturers wanted damages for loss of sales, especially manufacturers of condoms, the pill, baby foods, diapers, and such.

Some physicians were offended by lost revenues, especially psychiatrists, gynecologists, obstetricians, and pediatricians.

High prestige speakers would cover these topics. The Chief Justice of the US Supreme Court would speak. Two politically hot state Attorneys General would present papers and speak.

The Wall Street Journal offered speakers and a panel on the economic impacts of the virus. The Columbia School of Journalism offered a Simulated Problem on Virus-Related Legal Ethics.

The American Association of Attorneys presented the most popular sessions in the huge, enclosed, air-conditioned Ralph Nader University football pavilion. The attorney who secured a judgment of 2.3 billion dollars against the City of Los Angeles drew the most listeners. Los Angeles had failed to protect people from the aircraft which sprayed its citizens.

Meanwhile a "repair team" did maintenance on the air conditioning systems to assure that the cool air pumps functioned well in the enclosed arena.

All went according to schedule, and as the final speaker finished describing his lucrative strategy in the City of Los Angeles, the crowd applauded with great enthusiasm. At that moment, the athletic scoreboard began buzzing. It flashed the following message around and around the display panel:

To all you eunuchs, Esq., greetings. Take a deep breath. That clean smelling air is delivered to you by fellow citizens whose taxes paid for your awards, and it carries enough virus to assure that you won't have any more children to grow up to be lawyers.

An outraged roar burst from the audience, mostly male. The most agile men leaped to their feet and stormed toward the exits, managing to knock down and trample to death ninety-eight men and women, including several older attorneys and jurists.

Regis Jameson had arrived too late to get in the stadium and watched the carnage on TV monitors outside. He saw his fellow keepers of law and order climb over piles of bodies to escape. One man in back took several deep breaths, turned to his colleague and said, “Ah, what the hell. It’ll save me the price of a vasectomy.”

SIMULATED RENAL CALCULI

For several hours, Colleen and Jonah had comfortably slept in their Biosolz apartment. The room was dark, and when the bed rocked, Colleen dreamed she was back on the boat. A strip of tape was pressed over her mouth. Next her wrists were taped behind her and her ankles together. Hands slid her out of the bedroom window into a waiting van with Jonah just behind her. As the van moved out, she got a look at the hairless head and Mongolian features of a man staring intently at her.

How did they get into the compound? Colleen wondered. Mary bragged that we were safe.

The van accelerated steadily as they approached the security gate and broke through it. They drove a short time, slowed, turned sharply, and stopped. The door flew open and two men carried them into a dark hallway. After some muttering in what could have been Chinese, a dim light came on. Two men strapped Jonah to a bed. They inserted a needle into a vein on Jonah’s hand and hung an intravenous liter container set to fastest drip.

While they tied Colleen to a chair, she decided that they were more dangerous than the earlier kidnappers. They ignored everything she and Jonah did. Her fear increased as they pulled down Jonah’s pants. One of the men took a large surgical clamp and fastened it across Jonah’s penis. She burst into a muted scream, expecting blood to spurt. However, the clamp only prevented the flow of urine.

The man noted the time and sat down, staring at the bottle of fluid as it dripped into Jonah’s vein.

A fourth man came into the room with shaved head, immaculate hands, and a severe manner. He leaned over Jonah and spoke softly to him.

“You are familiar with the urogenital system. You know we have plugged a ureter?”

Jonah nodded warily. “Would you like me to loosen the clamp?” Jonah nodded quickly. The man did so, and turned to Colleen. “Watch your friend while I talk.”

Colleen fastened her eyes on Jonah’s. “You both are students of the conditions inside the human kidney. Correct?” Jonah nodded.

"You know that the only pain signals from the kidney occur when the fluid pressure in the kidney rises?"

Jonah nodded again.

"When the passage of urine is blocked, perhaps by a kidney stone, the pain becomes very great. You have observed this?" Jonah nodded assent. The man replaced the empty drip bottle with a full one while he talked.

"You will start to feel pain in both flanks. Soon, you will not be able to stand. You will sweat profusely. By this time the pain level will be beyond any other which you have experienced."

Pressing Jonah's bladder, the man said, "The ureter pressure is building, so I will apply the clamp. The final stage is reached when you vomit from the pain, and can only writhe helplessly until the clamp is removed." He turned to Colleen, nodding to the man behind her.

Colleen's tape was cut away from her lips. "Do not speak unless I tell you, or we will replace the tape. You are Colleen Jameson and you study at the Virus Laboratory at Berkeley?"

"Yes sir." The man nodded approval.

"You are a graduate student in virology and are studying reproduction physiology?"

"Yes sir."

"You have found a virus which causes the death of sperm after ejaculation?"

"Yes sir." Though untrue, she dared not qualify her answer.

"You work for the Biosolz Corporation. You offered large quantities of this virus to the People's Republic of China?"

"No sir."

He leaned forward and hissed, "Do not lie to me."

"We both work for Biosolz, but we only work in the laboratory. We developed the techniques to grow quantities of the virus. We know nothing of where the material goes."

"Have you heard any discussions that the spray might be used on Taiwan?"

"No. None."

"Have you heard of any planned military use of the virus?"

"No," Colleen said.

"As a biological warfare agent, how does this virus work?" he asked.

Colleen shook her head. "I don't see how it could be used for biological warfare. It doesn't disable people or make them sick."

"The news releases said that men infected with the virus felt pain for a day or two afterward. Is this not disabling?"

"I don't think so. I really don't know."

"You evade my questions," her questioner said. "Have you heard of the Spratly Islands?"

"No," Colleen said.

"Have you heard any discussions about oceanic petroleum or phosphorus explorations?"

"No," Colleen said.

"You know nothing of the conflict in the South China Sea between the PRC, Taiwan, and Viet Nam?"

She shook her head, "No," she said.

"You will regret lying to me." He nodded to the man behind her who taped her mouth.

In a few minutes, Jonah began to writhe, and he became beaded with sweat. He opened his eyes, terror stricken, and looked at Colleen. He knew the urine tubes in his abdomen would eventually burst.

The man leaned forward to Jonah and spoke.

"You are beginning to feel some pain. It becomes much worse." When Jonah retched, the man cut loose the tape from his mouth. Jonah heaved and groaned regularly. The man spoke to Jonah.

"You are working for the People's Republic of China? Tell me now and I will remove the clamp."

Jonah struggled to speak. "We work for Biosolz. We know nothing else." More retching.

"That's all you will say?"

Jonah gasped, "That's all I know."

"Fools," the man said. "Now you must come with us. You may not live through the next interrogation." He released the clamp from Jonah's penis and threw a towel in his lap. He spoke rapidly in Chinese to his aide.

The aide cut Jonah loose from the bed, and cut Colleen's bonds to the chair. "Get him ready to travel," he told Colleen. She nodded and rubbed her wrists to regain circulation. Jonah collapsed, a happy man, groaning in a puddle of urine.

Both their captors disappeared, and Colleen pulled the needle from Jonah's hand. "Get ready," she said. Jonah looked at her, dazed.

She loosened a rod from the intravenous stand and stood to one side of the door, her anger building. She heard someone returning, and as he stepped through the curtained door, she crushed his skull with the rod. He dropped without a sound.

"Jonah, get up." He struggled to stand upright. Colleen searched the downed man, found a revolver, and handed it to Jonah. The man who spoke English and interrogated Jonah returned. As he stepped through the door, Colleen hit him a glancing blow. He fell on the bed where Jonah taped him down as he, himself, had been. Jonah sealed his mouth.

Colleen motioned Jonah to watch the door. She placed the needle which had been in Jonah's hand into their captor's arm and started the drip bottle. Then she uncovered his penis and squeezed on the surgical clamp with all her strength. "Sleep well," she said as she turned from him.

"Let's go." They stepped out into the hallway and walked toward an exit door.

Another guard stuck his head out of a doorway. Colleen felled him with the bar. They moved quietly out the door and ran to the street. They heard shouting and saw two of their captors running toward them. Jonah aimed the pistol at them.

"I'm not going back there," he said.

A taxi slid to the curb. "Get in, quick," the driver said. He accelerated away from the curb and headed back to the Biosolz compound.

"Where did you come from," Colleen asked.

"Your boss has a couple dozen cabs out canvassing the neighborhood. I get a thousand bucks for finding you." The driver turned a moment to face them. "That's a smart lady. You know who I'm talking about?"

"Yeah, we know her," Colleen said.

In their quarters at Biosolz, they were checked by a physician, clucked over by Susan and Ted, tucked into bed, and shoved off to sleep with a sedative. A guard stood outside their bedroom.

The morning after Jonah and Colleen were kidnapped by the Taiwanese, Jonah awoke slowly from a nightmare. Jonah dreamed someone was fitting him with an artificial kidney stone.

"Is that you, Colleen."

"You'd better hope so," she said.

"I can't get it up this morning. No sample today."

"Now, now. Trust me. I'm a clinical specialist. I've handled a lot of pig penises."

Jonah was not reassured. "Maybe tomorrow. Stop. Ouch. Get your sample from Ted if you have to."

"Shush, Jonah darling. I was just checking to see if they did you serious damage."

"Did they?" Jonah stopped moving and listened carefully.

"Not that I can tell," Colleen said. "See, it's not too tender here, where they put the clamp, right? And it doesn't even hurt to put on the condom. Right?"

"No-oo," Jonah said.

Once again, she had her daily quality control specimen.

The Mob reassembled that night. All were filled with concern.

"What happened last night?" Ted asked.

Colleen told them until they made her stop.

"How did you get away?"

"They cut us loose to take us somewhere. Colleen knocked them down with part of the intravenous stand. We ran outside. I hobbled."

"How did Mary Henly find you?" Susan asked.

"She called in lots of cabs and with the security vehicles they all canvassed the neighborhood. A cab saw us as we broke out of the house."

"Lucky, you two." Ted and Susan spent the rest of the evening convincing Colleen and Jonah to sign the overseas contract. They found they were worried most about limited infrastructure, inadequate tools and equipment, untrained helpers, lack of freedom, and perhaps only limited access to the US.

The contract called for them to go overseas, to be under the protection of the national government of the country to which they were going. They would set up a production facility. Then they would train helpers to replace them, thereby making themselves superfluous. The host technical staff would operate the facility to assure they could do so after the Mob left.

Once a month $200,000 would appear in offshore accounts in Switzerland for each of them. Colleen would receive double the amount. When they signed the contract, they would receive tickets and a flight itinerary.

They decided to go.

23

CHINA

Once they signed the contract to leave Biosolz and go to China, the Mob ignored Mary Henly. They treated her politely, but she became merely the vehicle by which they would escape.

She catered an elaborate multi-course going-away dinner interspersed with toasts. They observed the toasts halfheartedly and returned to their excited speculation about the new job. Mary congratulated them on their hard work and achievements. But they only heard her crowing about the money they made for Biosolz. Colleen felt embarrassed but couldn't bring herself to reach out to Mary.

Jonah decided not to tell Colleen about the extension of the corporate spray plans to populations beyond Mexico City.

The Mob formed a pool to bet where they would go. Colleen guessed India because of the great need; Jonah selected Egypt because they had Western trained persons to influence government planning. Susan Trey chose Iran or Pakistan; Ted won because he always wanted to go to China. So much for logic, Jonah thought.

The first night in the People's Republic of China, the plane had hardly landed when the doors burst open. "Where are the scientists?"

A polite but very enthusiastic group of young, uniformed Chinese technical staff members joined the Americans.

A translator accompanied the technicians but was hardly needed. The technical members asked exactly the right questions:

"Where did you study?"

"What were your research interests when you found the active agent?"

"Is it a virus or a compound?"

"Which of you did what?"

"What is the active life of the agent?"

"Can you teach us to produce it?"

The attention enthralled the Americans. They were lionized while being paid to be there, and they felt flattered. They drank tea, nibbled on cookies, and talked technical matters for two hours.

After a uniformed figure appeared in the cabin, the plane soon emptied. As the technicians left they expressed great interest, gratitude, and desire for future contact. The Americans agreed. They seldom had such fun.

The uniformed man delivered them to primitive quarters and asked them to rest. He said they would meet highly placed political figures the next day.

The next morning, Susan went outside to see the surrounding rice paddies where people worked, bent over, hands in the water. The farmers planted the new rice shoots in water retained by a maze of retaining walls. She smelled the nightsoil and wondered if one got used to it. When Ted and Colleen came outside to join her, the guard became nervous and moved them back inside the building.

Eventually they went to another building, to a room where the military commander addressed them. A Chinese woman accompanying him spoke English with precision and a strong accent. The group understood most of what he said after a few exchanges with the translator.

"I am the officer responsible for moving your group and the group of technicians, with whom you talked last night, to your new location. It is a short trip from where we are at this moment."

"Before we join the other technicians, I must warn you about your discussions with that group. We in the Peoples' Republic of China wish this project to be a secret. We know that some persons inside China would stop it."

Jonah asked, "Why?"

"There is a strong tradition for large families in China. Couples in rural areas resist family planning. They must not know of this project."

"What would they do?" Colleen asked.

"I would not want to stir up the people who live nearby. For this reason, we will avoid discussion of your location while you are in the PRC."

"So we cannot travel during our stay?"

"That is correct. You should remain within the production compound until we are finished. With luck, you will finish in a few weeks. You agree?" He smiled and nodded at them.

They acquiesced.

"Last topic. Your mission is very important to the future well-being of the Peoples' Republic of China. The young military stewards of this group will act as guards. They are charged with your well-being and must keep close control over where you are and what you do. In their anxiety to keep you safe, they may . . ." He turned to the translator with a flurry of questions.

"Ah yes. They may 'talk tough?' But they may not touch you except to restrain you." He stopped in confusion. "They will protect you, although if they become frightened, they may make demands. Is what I said clear, so far?"

The group looked at each other and nodded. For once, Jonah didn't ask any smart-ass questions, and Colleen was glad he didn't.

"Our aircraft is here."

The boxy propeller-driven cargo plane carried military seats along each side. It had no windows from which to see and no insulation in the hull. Conversation was impossible. Jonah estimated the trip to be two hundred miles.

On landing, a bus took them to a cluster of two dozen buildings surrounded by a barbed-wire fence. They gathered in a large mess hall with tea servers on each table. A new uniformed agent spoke to the group through the translator.

"You all have co-workers you will meet shortly. They will take you to your quarters and tell you of the arrangements for food and housekeeping."

Four of the technical staff descended upon the group and walked with them to their quarters. They steered the group separately into four small rooms with bare bunks. Colleen's co-worker said, "I think you usually spend your sleeping hours with the man you call Jonah?"

"Yes," Colleen said.

Her co-worker took her into a small, spare room with a bunk, chair, and table. "It is small. It is all right?"

"It's fine," Colleen said.

After an awkward evening of fitting themselves into another small bunk, Jonah realized the honeymoon was over. He wanted his bed back, the king-size Biosolz luxury bed.

The almost military regimen swept them up immediately. Each morning, they rose quickly and ate a spare breakfast and a pot of hot water. They exercised alongside their co-workers. Then they threw themselves into converting a primitive lab building into a more modern, safely functioning virus production facility.

Ted Damkal told his partner Ms. Cho, "We are surprised that all of you speak English so well."

Ms. Cho seemed embarrassed. "We do not speak well. To be selected for this project, we must speak some English."

"So that you could assist us better?" Ted asked.

"So that we understand exactly what you say," Ms. Cho said. "We must learn to construct the laboratory without you present." Each of the co-workers will be required to rebuild the laboratory in another place.

COMMUNICATION

The differences in backgrounds demanded frequent training sessions among the Americans and the Chinese technicians. Colleen often worried that the technicians did not understand what she said. If they understood, did they have the vocabulary to answer?

The group used the first hour of each morning to orient and familiarize themselves with each other's technical vocabulary and accents. The discussions started with simple topics.

Sue asked, "Someone please describe the mechanism of infection by a virus." The group usually turned to Colleen as the authority figure until they located specialty skills in the various team members.

One of the Chinese technicians would struggle through the explanation: "Infection refers to the process by which a virus goes through the cell membrane and into the cell which it infects."

"But how does it get through the cell membrane?" Sue asked. Then came the drawings, words spelled, tentative explanations, and some tension and laughter, while the speakers tried to adjust their concepts and word selection to fit the groups' usage.

"Why does a virus cling to a receptor site on a cell?" Colleen asked.

Finally they managed, "The virus is like a plug that fits into the receptor site. The virus' electric charge pattern matches and clings to the pattern of the receptor sites."

"Then the cell is infected?" Sue asked. "Not yet. The cell membrane bulges inward, breaks down and the virus enters the cell. Now it has access to the nutrients and enzymes in the cell."

Colleen often pursued another topic. "If we are to produce large quantities of virus bearing solutions, how can we tell whether the viruses are alive in the solution?"

"Apply the solution to a pig?" a technician asked.

"That's one way, but we would soon run out of pigs. There is an easier way. Anyone remember the steps in sperm formation in humans?"

A male technician finally spoke. "The testes carry a coiled tube. The cells on the inside surface of the tube divide and produce the sperm cells."

"Good," Colleen said. "Now, do those tube cells have receptor sites, and how do you know?"

Most knew. "We know they have receptor sites because the virus enters the tube cells and destroys them. That is what makes the men sterile."

"Great," Colleen said. "Would you expect the sperm cells to have receptor sites?" she asked.

"When the tube cell divides and forms a sperm cell, the new sperm cell receives half of the cell membrane from the tube cell. So the new sperm cells should all carry the receptor site."

"Good." Colleen had them now. "So how do we test for live virus?" They thought for a moment.

"Put some active sperm in with the virus solution. If live virus are present they will enter through the receptor site and kill the sperm."

"Right," Colleen said. "The virus structure reproduces itself inside the sperm cell and uses up the cell's nutrients. The tail no longer whips"

The group appeared pleased to have talked their way through the problem.

"But where do you get the live sperm?" one of the Chinese women asked. Then her eyes went up with surprise. After a burst of exchanges in Chinese among the women, they glanced cautiously at the nearby males.

"You'll have to work out your own collection procedures," Colleen said.

The Americans often asked to be allowed to see the sights in the area surrounding the laboratory. At first the authorities refused.

After some weeks, Ms. Cho arranged a trip to the nearest city. A bus drove them for a few minutes before they reached the first buildings. Their co-workers escorted the Americans through the streets with one translator and guards ahead and behind the group.

The local citizens dressed much alike, a simple, two-piece, loose, cuffless blue pantsuit. They stared at the Americans. The group walked alongside an endless stream of bicycles, with bells jingling continuously. Workers rode in the streets on small, large, and multi-purpose bicycles of all variations. Many vehicles looked hand-made. Baskets woven to the bicycles carried produce, and children.

Colleen asked, "Who are those wearing masks?"

"Anyone with official duties," a co-worker said." They are responsible for cleaning the streets. That one is a policeman."

Unpainted wood surfaces covered the walls and walkways. Clothes hung to dry on every porch.

They arrived at a board fence which carried long pieces of printed paper. The citizens gathered around each strip and read eagerly.

"What is written on the sheets?" Jonah asked.

"This is our public newspaper," Ms. Cho said. "But not like your newspapers."

"How so?" Jonah asked.

"It is information released by the government. It tells about world events, gives news about production schedules, and who showed the highest output. Those are the graphs you can see. There is much urging to work hard and be responsible."

They came upon a large group performing Tai Chi.

Cho told them, "All Chinese workers take regular breaks for this. Also for eye exercises. If they are diligent, they will avoid the need for eyeglasses."

The Americans asked for a trip through the rice paddy area and the cooperative farm near their work site. They saw the simplest of huts and the few implements for cooking and eating. They saw the clinic for a cooperative farm, with its offering of free herbs, acupuncture and a few pills.

As the days wore on, Ted Damkal talked endlessly with Ms. Cho and slowly became enmeshed in lore of the orient.

At night, Jonah and Colleen huddled together, squeezed into their bunk, and whispered.

"Do you feel safe?" Jonah asked.

"Not really."

"What do you worry about?"

"A bunch of angry peasants breaking into the compound with agricultural tools," Colleen said. "Like the man said, we don't want them mad at us for producing the virus."

"We'd never get out alive," Jonah said.

Meanwhile, back in Berkeley, Colleen's professor, Dr. Shevlin, telephoned Biosolz Corporation and spoke to Mary Henly.

"I wanted to find out how my favorite student is doing, and whether you need any help from me?" he asked.

"Not yet," Mary said. "Ms. Jameson is out of the country on contract."

Shevlin snapped to attention in his chair. "Really? May I ask where?"

"The contract prevents me from saying. I can tell you that she has a very responsible appointment, lucrative too. We're pleased at how well she worked out."

"Well isn't that marvelous. I'm pleased. She was one of my best students," Shevlin said as they hung up.

Why don't I believe him, Mary Henly wondered.

Shevlin immediately telephoned his new friend, the San Francisco Bishop of the Holy Father Church.

"Have you had an opportunity to learn about the Colleen Jameson project?" Shevlin asked.

"I have, and it is a sordid story," the Bishop said.

"I am told that Ms. Jameson now holds a well-paid appointment overseas. She is doing research on the anti-reproduction virus and has new financial support," Shevlin said.

"In what country?" the worried Bishop asked.

"I'm not at liberty to say," Shevlin said. "I do hope she isn't putting the techniques of involuntary sterilization into practice in one of the European countries." That ought to scare him, Shevlin thought.

"Oh my," the Bishop said. He broke off to warn his superiors.

24

ADIOS, MUCHACHO MACHO

Once the Mob decided to go to China, Colleen called Regis.

"Mary Henly wants someone to represent us on the Mexico City Project. Jonah and I can't go."

"Why do any of us have to go?" Regis asked.

"If something goes wrong, they'll blame it on Biosolz." Colleen thought a moment. "I suppose that if all goes right, they'll also blame it on Biosolz."

So Regis Jameson found himself on a jet with project leader Steve Placer.

"Do you have any qualms about doing this job?" Regis asked.

Placer thought a moment. "I'm always uncertain in countries like Mexico. You have to rely on bribes to get yourself out of scrapes."

"Then you're not concerned about damaging the reproduction rights of these people?"

"Not really. In the U.S. military, I trained to kill people without worrying about their rights," Placer said.

"But we're not in the military," Regis said.

"Well, the Mexican President has sort of hired me as a mercenary soldier. Would you rather that I leave him with no way to reduce the suffering from overpopulation? Should some group approve the Project? A church? Or a political body?" Placer watched Regis.

"Well, I'm used to leaving 'justice' to decisions of a judge or jury. I guess the Mexican President feels he can't go that route."

They flew into Mexico City thirty days before the spray day, accompanied by the Spanish-speaking straw-boss.

They picked up vehicles and drove to the take-off site, a salt flat a few miles to the northeast of Mexico City. A nearby salt evaporation plant simmered in the heat.

"Ugly," Regis said, facing the city and the sultry smog and haze that extended over the whole area.

During the first week, the crew erected three tank shells with collapsible plastic liners and a long temporary shed. They graded and black-topped two roads to serve as landing strips with taxiways to the tank areas. They even put up a wind sock to orient the pilots.

Placer told the straw boss, "Paint big labels on the tanks near the valves, in Spanish and English." He wanted to make it easy for the tankers to get the fuel in the fuel tanks and the water in the water tanks. That kept the pilots happy.

During the next three weeks, they hired tankers to bring in water and gasoline. Next came the pilots Portakabins, the boxy European mobile homes. Steve Placer told the straw boss, "We're ready, maybe."

"Why just maybe?" Regis asked. He had noticed Placer's disorderly desk and worried about his tendency to make quick intuitive decisions.

"Well, I drew up all the flow diagrams we've been following so we wouldn't forget anything. But you're never sure until you test it." As a systems analyst, Placer considered every plan a failure until he field tested it.

"So how do you test it?" Regis asked.

Placer called Brownsville, the southernmost city in Texas, about 500 miles to the north.

"Ramon. We're ready for you. C'mon down."

"Where are you?"

"Down the coast to Tampico. Then head for the Mexico City international airport. You have to climb to 8000 feet. You'll see us near the evaporation plant, coiled like a snail. Give us a call." Steve gave Ramon GPS coordinates, the radio frequency on which they could talk, and their magnetic bearing from the International Airport.

The next day Ramon landed without calling in first.

"Any problems?" Placer asked.

"Naw. Good directions. And who could miss this big cloud of smog? Hijo!"

They loaded his spray tanks with water, filled him with fuel, and he took off to spray the nearby salt flats.

After four flights, Ramon steered the plane along the taxiway to Steve's truck. "What do you think?" Placer asked.

"Can you extend the take-off strips another 200 feet?"

"Easy," Placer said.

"Do it. It's hot and the air density is low up here. Hard to get off with a full load."

Placer then told Regis, "Now we're ready." He called Biosolz and ordered the virus. Tension began to mount. Four days to go.

The next morning, Regis saw a car drive toward them on the lake bed road. The car was an older model, carrying one man with a bushy mustache, an official look, a name tag, a clipboard with a pad of dog-eared pink forms, and a full key ring.

"What's going on here," he asked in Spanish.

The straw boss explained that they were testing experimental fertilizer sprays.

The man looked dubious. "I am with Airport security and they sent me to find out what you are doing. You must give me a copy of your contract and your permit to carry out the sprays."

When he heard they had no permits, he said, "There may be fines. Very expensive." He crossed his arms and stood his ground.

The straw boss sighed and took his hand from his pocket. He unobtrusively counted out three $20 bills, folded them, and handed them to the security guard as he shook hands.

"Thanks for coming out," he told the Mexican.

"I'll tell them you don't have a permit. Probably I will have to come back."

On the next day, the security guard brought his supervisor, and after some talk, he too accepted some money.

"You know that we must report this to our boss," the supervisor told Steve Placer.

That night, Steve and Regis talked at length about the various ways the authorities could stop them. Now, Regis was filled with doubts as to what he was doing there. What was the Spanish word for felony?

The next day, a strange looking plane arrived and landed without notice. Steve Placer ran out to tell the pilot he was lost.

The door opened, and Charlie Drummond hopped out. "Howdie Steve."

"Charlie? What the hell are you doing here?"

"How are you getting home when the festivities are over?" Drummond asked.

"I have a plane waiting in Tampico to take us to Houston."

"Well cancel it. We'll keep this one and take it home," Drummond said.

"Charlie, this is crazy. Things might get tight."

"Aw, let me stay, Steve," Drummond pleaded. "I don't get any fun any more."

Steve laughed. "Sure. What the hell. Glad to have you." Steve put his arm around Drummond and took him to a bunk in the Portakabin. Drummond looked around in delight.

"I spent my teens in a bunkhouse like this. What's going on now?" Drummond rubbed his hands, ready for action.

"We're painting the planes. When they arrive we give it a spray test run and then paint the word *Hombre* on the underwing. *Hombre* is the name of a commercial after-shave lotion. We paid for a radio advertisement that says '*Try the good smell of hombre.*' The local stations will play it all day when the planes fly."

"You want people to think the planes are spraying the aftershave lotion?" Drummond asked.

"Or maybe just advertising it," Placer said. "We paid a couple dozen taxis to drive around Mexico City with magnetic signs that say "*Buy hombre*," while they hand out free samples of *Hombre*."

"So it looks like an advertising blitz?"

“I hope so. It didn’t cost much, anyway.”

While the pilots painted the planes, another, more expensive car arrived with a man in a suit and two aides. Clearly brass, Regis decided.

After some discussion, the man in the suit told them they could not perform this commercial spray operation for *Hombre* Corporation without a license. He would send a police aircraft to assure that the operation had shut down. They promised to leave, wondering if they could get out before the police aircraft arrived.

With only 24 hours to go, a tank truck pulled up alongside the Portakabins. “Where do you want it?” the driver shouted. Drummond waved the tanker into the shed.

Inside, two technicians emerged from the cab.

“Where’s the water? And where do we deliver the mix?”

“Ready to go? Great. Over here.”

Steve explained the valve arrangements at the tanks. They mixed their first batch of virus spray, ready to load tomorrow’s planes. It was now 12 hours until spraying began.

That night, Regis and Drummond asked Placer questions about the project.

“So where did you get the planes?”

“I called three big agriculture corporations for lists of spray contractors. They told me who they hired, and I went to all the little guys they didn’t hire. These little guys jumped at the chance, hungry as hell.”

“Those are funny looking planes,” Drummond said. “You sure they’re any good?”

“Piper Pawnees. They make up most of the world’s ag planes. Finest kind,” assured Steve. “Also the safest.”

Regis wondered why he would worry about pilot safety when he ignored damage to the Mexican civilians.

The pilots worked out a traffic pattern to avoid mid-air collisions. They also planned their spray pathways to try and avoid complaints from the Air Traffic Controllers at the International Airport on the South side of the city.

Ramon came to Steve. “I grew up over there.” He motioned toward the smog cloud. “Mind if I go first, tomorrow?”

“Sure,” Steve said.

Regis listened in awe as Steve Placer briefed the pilots on the mission.

"This experimental program delivers a chemical that precipitates out the solid particles in the air over the city and reduces damage to people's lungs."

Ramon grew up in Mexico City and wanted to do something for the people of Mexico. After all, they caused him to emigrate to the US, where he learned to fly. He took off at dawn, like in the old movies, and started his spray at the first cluster of houses. The Basilica de Guadalupe was the first structure to catch his eye; then he passed the Colegio Militar. Just as Steve Placer demanded, he avoided the Zoological Garden of Chapultepec which housed primates. He dodged the airport and headed back to the flight strip, disappointed when he saw just as much smog remaining.

At ten a.m., an observer arrived, paid to bring news from downtown.

"Anyone react to the airplanes?" Steve asked.

"No. The people ignored them."

"No comments on the reason for the overflights?"

"None at all. Well, a few little kids pointed."

"Did you hear the radio advertisements?"

"Yes. Sounded like normal for the radio." No, he didn't see any taxis with signs. Steve sent him on his way.

At that point, a light plane marked "Police– Federal District of Mexico" began making passes over the flight strip, dodging spray planes in the landing pattern.

Steve Placer and Charlie Drummond ran to the plane and talked to the passenger. He was angry and assured them he would be back on the ground. He would bring sufficient men to stop them and arrest the responsible parties. Drummond ran to warn his airplane pilot, and Placer ran to the Biosolz spray pump operator.

"Shut her down." Placer found the driver of the tank truck.

"Head back along the route you came, and keep going until you get to Brownsville. Then you are on your own. Don't screw around down here. You'll get in trouble and lose the rig. OK?" He handed the driver a bonus payment.

"You got it, chief. I'd still be married if I'd driven for you all these years. I'd have been home on time." You're not home yet, thought Placer. The driver found the technicians and hurried them into the cab.

Each pilot showered while someone refueled his plane. They

changed their own oil, and got their bonuses. Steve sent them off with instructions;

"Remember, you have to get over the border into the US to collect the final payment. Here's the phone number. Don't take any drugs back. Not even one little joint. You aren't smugglers."

They nodded seriously, stuffed the money in their pockets, and took off to the north, heading for the Rio Grande and the flight strip in Texas.

"Gawd almighty," Charlie Drummond told them. "Theres a plume of dust behind some trucks goin' top speed, turning into our road."

Regis, Drummond, Placer and the straw boss ran for Charlie Drummond's plane,

Placer looked at the angular high wing craft and said, "I hope it's good at short take offs."

"S'posed to be," Drummond said as they tumbled in. The pilot started the engine, taxied out, and was off the end of the runway when the first police truck arrived.

They all made it home without incident except for one of the ag pilots. He said he had engine trouble and barely got out of Tampico two days later. Regis decided that Steve Placer's speech about drugs reminded the pilot how to improve his earnings for the trip.

25

POOR RELATIONS

A dozen environmentalists (people the commercial press called "eco-terrorists") worried about the new virus and met at a camp-ground outside of Reno, Nevada. Biologists and marine scientists led the organizations Law of the Sea, and GreenSea, and published the environmental journals *Wild Country* and *Earth Prime*!

Some called the sailors of the Law of the Sea "mariners with a death wish." From a small ship, they disrupted drift fishermen operating in violation of international treaties. They rammed illegal fishing boats, picked up illegal nets, and asked the fishing skippers to meet them ashore to make a complaint to the authorities. The illegal fisherman seldom came to shore, not wanting the publicity. GreenSea especially protected threatened marine mammals and prepared devastating television documentaries on pollution.

The *Wild Country* published warnings of worried conservation biologists on how much damage the earth's ecosystem had sustained. *Earth Prime!* printed news for lovers of forest ecosystems. This news consisted of screams of agony and rage from those who watched the last stands of timber cut and wildlife pushed out of habitats.

Everyone respected Paul Watchman for his leadership and fearlessness in confronting the drift net ocean pillagers. He had called them together and started the discussions. "I guess you all followed the virus stories. I have the *Wall Street Journal* article if you want to see it." They nodded somberly.

"Anyone know Colleen Jameson? Should we try to talk to her?"

One listener spoke. "We asked around, and she appears to be in hiding. She was a grad student at Berkeley, found the virus, and applied for the Birnham Prize without thinking what might follow. She may be a technocrat with no commitment to conservation."

"Shit," someone said.

"How about the Berkeley Microbiology Lab?" someone asked. "Should we consider action against the Lab facility?" Several expressed enthusiasm for this.

"If it would fix the problem, I'd go for it," a GreenSea member said. "I've been in the lab, and I think it's vulnerable. However, the *Journal* article pointed out that the Birnham Foundation already received applications. I think the virus is already disseminated."

"Shutting down the lab won't stop the virus from getting out?" The *Earth Prime*! enthusiast looked dismayed.

"Right."

"What about approaching the Birnham Foundation? Ask them to stop the Prize?" suggested a *Wild Country* writer.

"Call them right now," they agreed. The best known research ecologist agreed to call. "Talk good," someone shouted.

Watchman suggested they brainstorm a list of bad effects of a world-wide virus spray. "Well, if the agent causes permanent damage, it's more dangerous."

"Is it communicable from the exposed person to others?"

"Yeah, hard to stop that. We gotta hope it isn't that bad."

"Then, depending on the characteristics of the virus, we might lose the whole human race?"

"Maybe."

After a long silence, someone said quietly, "Well, then, things aren't all that bad."

The group burst into laughter. They quieted down for a bit, then someone started to laugh again, and they all laughed until they rolled on the ground. Someone said, "That does it," and brought out a case of beer. Some rolled cigarettes.

They rearranged themselves, with an occasional outbreak of giggles, and addressed their fears.

"How about other animals?" Watchman asked.

"Yeah, there's the bad shit. We could expect our most closely related species to be in trouble. At worst, a really toxic chemical agent could extend far down the family tree."

"I'm not that good a biologist. Tell me who might be jeopardized?"

"Well, all the other races of humans, the apes, monkeys, lemurs, tarsiers, and other primates. It might threaten the mammals including porpoises and whales.

"But that's nearly everything, all the so-called 'higher forms,'" a woman complained

"Well, the plants will have a heyday with all of us for fertilizer."

"How about the social effects on large groups of people who lose their reproductive ability?" They turned to their sociologist.

"Some environmental sociologists have made predictions. If a certain racial or ethnic group can attack another and destroy its reproductive potential, then genocide is likely. Many possibilities come to mind."

A truck approached with their emissary to the Birnham Foundation. He settled in, beer in hand, and began.

"At first they didn't want to put me through to anyone. When I told her who I represented, the receptionist gave me another number, and said that Jonathan Remmer, PhD, would probably speak with us."

"I called the number, and here is the conversation. It's easier if I just play the tape of what he said." He turned on the player.

"This is Jonathan Remmer. I understand you are calling for some environmental groups. What can I do for you?" Remmer asked.

"We are quite worried over any widespread use of a virus which affects reproduction."

"I imagine you're worried over side effects of the reproductive spray," Remmer said. "We debated that for quite a while."

"We are worried about the impact on our closest animal relatives."

"Certainly," Remmer said. "That possibility gave me the greatest anguish of all. We mapped the present natural colonies of primates which might be threatened. I believe most are at considerable distance from the high density human groups which may receive the largest amount of the virus. But certainly your worry is legitimate."

"Is there a chance that we might talk you out of the plan? We would like to meet with you," the ecologist said.

"I'll be glad to meet with you. However, I must point out that the scenario which we predicted is already in progress. It's really out of our hands," Remmer said.

"He'll arrive on the morning jet into Reno," the ecologist said, stopping the tape.

The next morning was cool. They gathered around the breakfast campfire, with coffee cups.

"Let me tell you the scenario we predicted," Remmer said, "and tell you how far it has gone."

He talked through the day. They interrupted often and contradicted him, hopefully. He described the expectations of the Foundation group which offered the prize. The listeners joined in to try and rewrite the scenario. Usually they fell into Remmer's logic.

By mid-afternoon, Remmer slowed down.

"So, dear friends, even if we at the Foundation have miscalculated, we are committed. The events have gone too far to overturn. I talked to the brother of Colleen Jameson, the graduate student who discovered the virus. She has not the vaguest notion of what will happen. Nor have we. From now on, we can only hope. With any luck, we might be able to intervene occasionally to reduce the damage."

"For better or worse, the virus only attacks male fertility. You recall Hardin's warning that women's loss of fertility would bring about much greater suppression of population than men's loss."

Remmer suddenly spoke very forcefully.

"More than anyone else, you people are more aware of the catastrophic environmental effects caused by human overpopulation. Think now. If you could reverse what has happened and take away the virus as a possible population control device, would you do so?" Remmer sat and watched them closely.

The group fidgeted, some shaking their heads, caught between the ends and means of the environmentalist dilemma. Watchman finally said, "Maybe not." He asked for any last words as Remmer left.

The ecologist began. "I started out the day in a fury. Then we all discovered our mixed feelings. Now, I feel a bit more hopeful. I really thought the higher mammals might die out."

The others had similar feelings.

They thanked Remmer for coming. Some wept with fear for the future. Remmer saw their dismay. No one worries like tree huggers.

26

STEP ON A SCORPION

The Iranian oil magnate who bought the first batch of the Jameson virus at the Northby auction took it home to Teheran. To find and outfit an airplane with tanks and spray equipment took several more weeks. The Iranians found an aged DC-3 which they hoped would frighten no one, not even the Israelis. However, they took the precaution of painting Israeli markings on the outside.

After installing four drums of virus, the Iranians recruited two Iraqi pilots to fly the mission. They paid the pilots very well and stripped the plane of any Iranian identification.

Finally, the Iranians negotiated arrangements to allow the plane access to Israel through the North. The pilots planned low-level passes over Haifa and Tel Aviv, as long as their fuel and cover held out. When the Iranians told the Iraqi pilots the spray was a poisonous gas, the pilots laughed and congratulated themselves over the opportunity to strike at Israel.

After the first leg, the pilots refueled at Baghdad, Iraq, and tucked the gas receipts in their pocket. They started spraying soon after the plane crossed the Lebanese border. Following the coastline south they released short bursts over the smaller communities. They used up two full drums over the Bay of Haifa area. By the time they reached Tel Aviv, some seventy miles south from where they started, anti-aircraft devices warned of an aircraft approaching. The spray-plane emptied its tanks by the time fighters shot it down.

The Israeli authorities examined the wreckage and quickly discovered the Iraqi documents on the plane and pilots.

Iranian agents planted stories that the Iraqis had spread a dreadful "castration gas" over Israel.

The UN reacted predictably. "Saddam Hussein and his military government has again gone too far. Why did they illegally send aircraft over Israel for some unspecified purpose? We warn him once more that he must stop this aggressive behavior toward his neighbors."

As stories describing the aerosol spray appeared, the UN challenges to Iraq became more belligerent. "We will take action against Hussein and his genocidal followers if he continues these provocations."

Iraq denied any knowledge of the operation.

During the next few days, epidemiologists from the World Health Organization helped Israeli doctors examine the citizens. Their sampling predicted that among the Jews in Israel, the spray sterilized as many as a quarter of a million males. When the press announced the news, some civilians went into mourning and others rioted.

Palestinian and Arab groups discounted the claims, saying the UN usually sided with Israel. The world's Arab communities churned with amusement and vulgar stories about Israelis and their problems.

Sixteen hours after the announcement of the sterilization of the Israeli citizens, an aide handed the seventy two year old President of Israel a sheet of paper. He read the message to his close advisers, who waited with him for this announcement:

At thirteen minutes past noon, local time, Israel, without warning, dropped a medium yield nuclear weapon on Baghdad, Iraq. It exploded at low altitude. The fireball touched the surface and carried up large quantities of earth that continues to rain down upon the surrounding area. The black cloud obscured the sun. The perimeter of the city burned actively. Over four million people inhabited Baghdad.

The President of Israel nodded, his mouth set in a grim line.

"Let those paralyzed hypocrites at the United Nations cry their tears upon us. Can they wash away our loss? They did nothing, and we did something. These jackals will think twice before they attack us again. It is we who must bear the loss of our children and grandchildren."

His advisers muttered their infuriated agreement. "They criticized us after the Holocaust, and told us we should have fought the Nazis," one adviser said. "Then they cheered us when we fought the Egyptians. Now, they complain when we fight the Iraqis."

Israel's President wondered why he happened to visit Haifa on that terrible day. His fury broke out again.

"If they tread upon a scorpion, they must expect to suffer." Growling, he shifted in his chair trying to ease the dull pain in his groin. He wondered what his wife would say.

Those residents of Baghdad not killed outright soon succumbed to lethal levels of radiation from radioactive dirt and particles settling on the area.

Further out, those who did not die of radiation poisoning during the first few weeks could expect that their children and their children's children for five and more generations would show bodily damage from genetic mutations induced earlier. These changes might be gross deformities, or more subtle changes and functional losses caused by the ionizing radiation received from the long-lived secondarily radioactive fragments around them.

Israeli supporters in many places privately supported the bombing. So did the high-echelon Iranians, who planned and carried out the spraying of Israel while publicly criticizing the Israeli bombing of Iraq. Iranian Shia Moslems struck a blow against the Jews in Israel and caused the retaliation to go against Iran's old enemies, the Sunni Moslems in Iraq.

The Americans noted that Saddam Hussein finally died. The Iranians accomplished what the vaunted Operation Desert Storm had failed to do. Hussein's death was brought about by a combination of stealth, deception, and power, in the tradition of the earlier Persian and Byzantine Empires.

In San Francisco, Regis decided not to tell Colleen and Jonah. Already worried about their stress level in China, he didn't want to add to it.

27

DISSENT

The Americans in China had to face their major equipment shortage. Jonah raised the problem with his boss.

"We need tanks, Colleen. Nice new shiny clean stainless steel tanks. How many, and what kind can we get?"

Colleen asked her co-worker, but no one could help until they asked Cho.

"Wait," Cho said. She left the group and returned shortly with drawings. "The airplane can bring us these sizes in about a week," she said.

It turned out that the Chinese gave Ms. Cho the most authority in the group. Half Korean, she stood a bit taller than the others. She concealed from the Americans her Master's degree in microbiology, and always deferred to Colleen and the other visitors.

"We've ordered the tanks, and I"m ready to draw the flow diagrams, but how we build the production system depends on the pipes, gauges, and valves we can get."

Cho solved part of the problem. "In a few days, they are sending us a welder and a supply of stainless piping. We also found a steam generator in the city near here. It should arrive tomorrow."

Jonah said, "I'm most comfortable assembling the physical system, like Ted and I did at Biosolz."

But Ted wanted a change. "I want to get away from the grunt work. How about if I help Colleen with the system design?"

Sue tip-toed into Colleen and Jonah's room one morning. "News," she said. "Last night, I saw Ms. Cho disappear with Ted into his room. Then I saw her leave this morning."

Jonah groaned. "Some call that 'fraternizing.' I hope our new organizational mix is congenial. I'd hate for Ted to have to go back to grunt work."

"Now, Jonah. I'm charmed," Colleen said. "Why do you always expect the worst?"

Because that's what usually happens, Jonah thought.

"I wonder if Ted found her as attractive before we knew she was our boss?" Sue asked.

"You two cynics," Colleen said.

When she got Ms. Cho aside, Colleen asked, "Did you get in trouble because you stayed in Ted's room?"

"Yes," Cho said. "They said I disgraced the PRC because of this relationship. They threatened to remove me from the project."

"What did you do?" asked Colleen.

"I told them I would break it off immediately."

"You broke it off?"

"No," Cho said. "They told me to continue, and not disrupt the project."

Colleen burst out laughing. "Here we are, virus technologists, into the 21st century, and we're still chattel. Nothing's changed." Even Cho laughed.

Construction went more rapidly after the welder arrived.

The Chinese contract with Biosolz required that the Americans teach the Chinese each step of the virus growth process so they could design future plants. Ted Damkal taught the session on the Biosolz growth process to the Chinese technicians. He asked a lot of questions.

"Most researchers grow virus cultures in primary cell cultures or fetal cells. Why won't that work for us?" Ted asked.

"After those cells divide a few times," Cho said, "we would have to bring in new ones."

"Right," Ted said. "So, we chose a continuous cell line. Who can tell us about these?"

"I think Ted is coaching Cho at night," Jonah whispered to Colleen. She pinched him silent.

"Continuous cell lines are malignant cells which have been taken from a human," Cho said. "These cancer cells divide rapidly and repeatedly, without aging."

"Right," Ted said. "So we can continuously grow virus in our cancer cell cultures."

"But don't we need cells from the seminiferous tube present?" a technician asked.

"Yes," Ted said. "We started a continuous line of cells using malignant tissue from pig testicles. We got the tissue from slaughterhouses."

"And that is the culture medium in which the virus grows?" asked a technician.

"Yes," Ted said. "They are different from normal cells. They have varying numbers of chromosomes and simpler nutrition requirements. We'll grow them in their own separate tanks, and harvest them by taking batches of the solution. By itself, the virus can't grow. It needs these cells."

"Where will we get this cell culture?"

"We'll have a quantity sent from the United States when we are ready. After that, it will be your job to keep the culture medium healthy."

"Won't the cell culture be a good growth medium for stray bacteria?" Cho asked. "How do you keep it free of infection?"

"You're right," Ted said. "It will support bacterial growth that could kill the cells. We must use sterile handling procedures. If necessary, we will also use antibiotics to kill bacteria."

"Will we also receive a sample of the virus to be grown?"

"Yes, we'll have that flown in. We'll prepare some growth tanks for batches of the culture cell medium and add small amounts of the virus." Ted asked one of the technicians to tell what happened next.

"The virus will come in contact with the culture cells' receptor sites and enter the cells. Then the viruses react with the enzymes and cell nutrients producing proteins which are assembled into new virus structures."

"Great. Keep going," Ted persisted.

"The cells fill up with viruses. At some point, the cell membrane breaks and frees the newly formed virus. This increases the number of virus molecules in the culture medium."

"Right," Ted said. "That increase is the reason for the whole project. Then we transfer the fluid to storage tanks and chill it."

When the first virus samples arrived, guards escorted them into the laboratory. All gathered around for the crucial event. Sue performed the glove box transfer operation. Three tanks awaited inoculation from the dozen transfer bottles sent by Biosolz. The first two transfers went smoothly.

As Sue's co-worker reached into the Biosolz container, she withdrew the bottle, and moved toward the entry port of Sue's glove box. Suddenly she twisted the lid, breaking the sterile seal, and threw the contents of the bottle into Jonah's face as he stood watching.

She screamed some words, pulled a screwdriver from her pocket and leaped toward Colleen with her arm raised.

Colleen stood horror-stricken as she realized that Jonah had been contaminated. The raging technician stabbed her in the shoulder.

The guards leaped in and dragged the screaming woman out of the laboratory.

Ted Damkal quickly fitted a respirator over his nose and mouth.

Colleen's wound was superficial. However, the virus infected Jonah and two guards. That night, Jonah's testicles ached as the seminiferous tube cells swelled. The next day, after the cells burst, the tubes scarred and became sealed.

Jonah withdrew from everyone. Colleen tried in vain to get him to talk about what he was feeling. When she put her arms around him he pulled away. He barely spoke to anyone and became more and more distant.

28

SPLITTING UP

Seven days after the Chinese technician attacked Colleen and Jonah, the Chinese authorities called the Americans into the assembly room, Colleen with her arm in a sling. A uniformed guard stood by. The translator spoke to the Americans about the attack.

"As representatives of the People's Republic of China, we offer our apologies for harm done to you. We investigated how the traitor managed to insert herself into the technical staff group. Her credentials seemed authentic. We have requested additional guards and several new technical staff members."

Ted asked, "What qualifications will the new members have?"

"We do not know. Probably training in one of the biological sciences. These people are in demand, and we must take them from other facilities."

Colleen asked, "Do you know why she attacked us?"

"We do not," answered the translator after some discussion among the Chinese.

"What did she say when you questioned her."

After more discussion, the translator said, "She was vague and evasive."

Just like you, Colleen thought.

Jonah still would not speak to anyone.

The translator tried to finish the session, but Colleen spoke out again.

"Surely you must have some idea about why she did it. She acted as a political agent? She represents a religious group?" Colleen thought to herself, I'm still angry about my cat.

Again, after more discussion in Chinese, the translator said, "We do not know." Colleen persisted.

"It is very frightening to be attacked and wonder if it can happen again. Is it possible that other technical staff members plan to attack us?"

"No," the translator answered immediately. "We questioned

them." With that, the translator said, "This meeting is over. Please continue your work. The other staff members will return soon." The Chinese delegation then left the room. The guard escorted the Americans back to the laboratory.

During the next few days, the Chinese technical staff members returned, one at a time. Ms. Cho arrived last. She seemed calm, but she told them nothing.

Ted complained to Sue that Cho remained at some distance from him. "You would have thought Cho would tell me something," Ted speculated. "I guess she feels I might not believe her."

"Excuse me for asking," Sue said. "Its my anxiety. Did she let you start your relationship again?"

"You mean sex?" Ted asked. "Yes, but her heart's not in it. It's as if they told her, 'Do what you have to.'" He added bitterly, "Anything to keep up the production level."

A few days later, a new authority figure arrived on the scene. They had been in China for six weeks, although it seemed less. Susan Trey, however, marked off 42 days on her wall. The newcomer called the American group into a consultation room.

"My name is Chloris van Damme. I am now the political project leader for your work. My Dutch father and my Chinese mother reared me in England. You should probably call me 'Damme.'" She seemed hesitant and distracted, pushing papers around in front of her.

"The People's Republic of China would like to congratulate you on your accomplishments to date. As concrete notice of our pleasure, I bring deposit slips for the sixth week of service, along with the expressions of gratitude given to me by the political leaders of the Peoples' Republic of China."

The group peered closely at the Swiss Bank account deposit slips, fearful that so many zeroes indicated a mistake or a misprint. However, they saw Colleen's brother Regis' imprimatur stamped clearly on each slip, verifying the deposit. With their feelings buoyed, they listened with renewed interest.

Damme went to the door and brought in the three remaining co-workers. Now, Damme addressed the senior technical group of seven.

"We wish to set in motion the second portion of the contract. We wish to replicate what you have accomplished at this facility in another part of the country. We will start dispersing the product of this facility soon, and wish to begin construction of other production sites." Damme pressed on.

"Could you give me some idea who might like to go to the new site to head up construction, training, and production?"

The group turned to Colleen and waited.

"I would prefer to stay here. So each of you should speak for yourself," Colleen said.

Ted turned to the group, almost to seek their permission. "I would like to go to this new project. It is an opportunity." His eyes blazed. "Especially if my co-worker Cho could go as well. We are working closely together, and I would like her assistance."

"Good, good." Damme said. She turned to Cho and said, "You agree?"

Cho bowed immediately, obediently to Damme, and said, "I thank you for your permission. It will be an honor."

"Does anyone else wish to speak?"

Susan Trey hesitated, then spoke up. "I'd like to help set up the new facility without depending on Colleen and Jonah. That would make me feel more confident."

Damme said quickly, "Very good. I hoped you would agree." She turned to the other Chinese technician.

"You are willing to go to a new site?"

"It is my honor," she said.

"Very good. Thank you." Damme wrote a note, then turned to Jonah.

He looked straight ahead. Finally he said quietly, "I will tell you tomorrow." Colleen stared at him.

"Very well." Damme looked around the group. "Then observe your goodbyes with each other. We do not know exactly when you will depart. Once again, you won't know where the others are."

Colleen stumbled through the rest of the day, her face rigid with apprehension.

At dinner, Ted and Susan tried to thank Colleen for the gifts they received, of information, money, friendship. Colleen tried to sound enthusiastic about the support Ted and Susan gave them at Biosolz. Jonah left the room.

The new assignment transformed Ted. His eyes gleamed as he thanked all for their friendship and help. He saw a new world to conquer with Cho at his side.

Susan looked to the future, positive, but subdued. She pleaded that they all meet in the near future. She suggested that they keep track through Regis.

"Of course," they assured each other. After their tumultuous time together, no one really believed it could end.

Colleen broke away and went to her quarters. Jonah lay on the bed.

"We have to talk," Colleen said.

Jonah nodded.

"Are you leaving?" Colleen asked.

"I don't know," Jonah said.

"Oh, Jonah," she moaned. "I'll die without you."

"No you won't. You're tough. I've been gone since the attack," Jonah said.

Colleen drew near and sat on the edge of the bed. "Once you promised to always tell me what you are feeling."

After a long silence Colleen thought he withdrew again. Then he started speaking slowly.

"I don't know what happened. I thought that I followed the rules, studying, being responsible, trying to help people."

She reached out to put her hand on his shoulder. He shook it off. "You can't help, and I wish you could. The last thing that I remember, I got honors on my exams and worked on my doctorate. Then I met you."

She froze in terror of his next words.

"That was the best of all."

Colleen sobbed with relief. Jonah went on.

"Then they killed Bearing. They killed your cat. They tried to kill me, but you stopped them." He glanced at her. "I'm not sure I ever thanked you."

This time he didn't shake off her hand.

"Suddenly, I'm not an honors graduate student. I may be a fugitive from the law. How did that happen?"

After another pause, he continued. "I thought we fooled them. I hoped we were safe, hiding in the arms of corporate America, behind a fence, with lots of money to rub on our wounds." He turned over on his back and looked at her.

"We sure fooled them. Except when some crazed Taiwanese kidnapped us and tried to practice reverse medicine on me."

His eyes had a bit of light in them. Colleen lay down with her head on his chest, crying quietly.

"Then they poured on more money, and said we had to leave the country and come to this struggling insane asylum where we try to stop too many people from dying in each others' cookstove smoke." He sighed.

"You're getting me all wet," he said. She nodded.

"What's our reward, this time?" he asked. "They stabbed you, and I'm so sorry. I can't stand to have you hurt." She started kissing his cheek.

"And of course, whatever they did to me."

"But I don't care," Colleen said. "I love you. I can't bear to watch you suffer."

"I know. You are the last thing I have that I care about."

"I missed you," Colleen said.

He wiped her face dry. "You're a mess. I missed you."

They finally made love, timidly, as if for the first time.

"I got lost," he said.

Colleen didn't care. He was back.

The next day, Jonah told Damme, "I'm staying."

"Very good," she said.

"A letter from Regis." Colleen read quickly. "Remember the speech he sent for our opinion?"

"For the National Association of Riflemen," Jonah said. "They should join up with the Wildness Club?"

"That's the one. He gave the speech in Las Vegas, to a big group." She paused. "I'll be damned."

"They rioted?" Jonah asked. "They shot him with the National Association's official rifle?"

"No," Colleen said. "They laughed. They booed. They argued. Someone threw a chair onstage."

"Then came the tar and feathers?" Jonah asked.

"No," Colleen said. "They voted, and approved his plan, 71 to 41, and nobody quit or walked out."

"I want to see a copy of that speech. He's pulled off a coup," Jonah said. "Amazing."

29

DECISIVE POLITICAL ACTION

The President of the United States held a meeting with his Secretary of Health to consider the virus problem and what the government should do.

As they spread papers on the table, the President asked, "Are you still mad at me, Phil?"

The Secretary looked up, wide-eyed. "What about?"

"You know. The Mexico City visit. I pushed you pretty hard to go."

"Oh, that. It's a good thing you did. I needed the reality shock. At first, the Consul refused to take us around. When we made the tour, my wife went too."

"So what happened?"

The Health Secretary noticed that the President leaned in and listened closely for his answer. "Well, we saw and smelled the worst of Mexico City for a whole day. We talked to the local health officials for quite a while. Then we caved in and hid in our hotel. My wife and I have been talking about it since."

"Do you have any ideas for solving the problem?"

"No."

"I don't either," the President said.

"I have to thank you for forcing me to go down there. My wife has a whole new respect for me, and for you. Since we went, she asks me about my job."

"Great," the President said. "Now, what did you find out?"

"The most helpful ideas came from an old-timer from the Third World public health wars. He's in the UN World Health Organization and gave me several hours of his time." The Secretary shook his head. "Basically, here's what he said."

"You can't help a population if it is growing faster than the food supply. Since 1992, we've had reduced amounts of food per person. If you save people from starving this year, they come back with bigger families next year."

"So our emergency food exports don't help?" the President asked.

"No, they make things worse. Of course if you say this in public, religious groups attack you for being heartless and anti-family."

"So the virus sprayers are really helping relieve long-range suffering?" the President asked quietly.

"They are if the population drops somewhat in those areas," the Secretary of Health said.

"So what do we do with public demands for action against the virus sprayers?" the President asked.

"Be vague, I suppose," the Secretary said. "Remind the public that the contractors are based outside of the United States. Talk about elusive transnational corporations, global reach, jurisdiction problems. Also, tell them that several countries offered money for spray programs to be applied to their own people."

The President frowned. "But isn't this just the wealthy people reducing the number of poor people in the world?"

"Yes, it is," the Secretary said. "And we have to do it until the population is down to a quarter or so of what we have now. If we don't do it with a spray, it will happen due to deaths through starvation and competition for food and most likely wars and epidemics, like AIDS or Ebola. Maybe worse if we let them cut back the rainforests any further."

"But shouldn't we also be reducing the number of children that the wealthy have?" The President leaned forward, very keen for his answer.

"Absolutely. In fact, even more so. The Carrying Capacity Foundation showed that one wealthy child uses up perhaps 200 times the resources that a child from a poor country uses."

The President finished his argument. "So if a parent in the U.S. chooses not to have a baby, they have just reduced environmental damage 200 times compared to a poor woman who avoids having a child."

"That's right."

After a pause, the President said, "In that case, shouldn't we try to get some spray onto our biggest cities?"

The Secretary of Health laughed. "It would save a lot of starvation in the near future. Oh, I forgot. Elquart Pharmaceuticals says they have a vaccine for the virus, available in under a year."

"Don't tell me," the President said. "It'll be too expensive and hard to apply to the world's poor. But the Congress will underwrite U.S. distribution through the Public Health Department? The wealthy win again?"

"Your prediction sounds very likely. You would probably do well in politics, sir."

The President rose and shook the Secretary's hand goodbye. "I doubt it," he said.

KIN TAMA, LTD., A TRANSNATIONAL SPRAY CORPORATION

Flavin Taylor hosted a progress meeting in Houston. He picked up Charlie Drummond at the airport. Drummond told him how much fun Flavin missed at the air strip when they sprayed Mexico City. In the elevator, he told Flavin that he had figured out how to thank him for the good time he had visiting Flavin and his wife at Hilton Head.

"How?" Taylor asked, warily.

"I asked your boss in London to assign you to our next field operations. You can go out with me," Drummond said, beaming.

"No." Taylor said. "I don't want to go. You are just making sport of me, right, Charlie?"

"Naw," Drummond said. "Your boss thought it would be a great idea. You'll have a ball. I'll let you know when we leave." Taylor watched Drummond walk away, his apprehension turning to fear.

Regis Jameson flew to the Houston meeting to represent Biosolz Corporation. He sat quietly toward the back of the conference room

The Detroit Management Group arrived and described the new corporation, Kin Tama, Ltd., and its headquarters in Caracas. The President of Kin Tama was a Japanese who spoke Spanish, on loan from Holland Petroleum.

As Purchasing Agent, Steve Placer would oversee the corporate operation from Houston. Then Placer asked Charlie Drummond to report on the Mexico City spray operation.

Drummond leaped to his feet, waved his arms, and talked rapidly. By the time he finished, his eyes glistened, his hair was in disarray, and he wanted to go on, but the story was over.

One of the Britts looked carefully at Drummond. "It appears as if Charlie should take a leave from the bank and head up the next field operation. What do you think, Steve?"

Steve nodded. "Absolutely." Flavin Taylor moaned.

Drummond lit up. "I'll do it. What do you think, fellas? Really?"

"Just let me know," Steve Placer said, "and we'll plug you in, Charlie. You did good."

Drummond sat back, excited, and pleased.

Steve Placer flipped on an overhead projector. "Now for future business. The chart shows what we have arranged for bid sharing:"

Target:	*Bidders:*	*Bid in mil:*
Mexico City	Mexican Govt.	$19
Japan Islands	Detroit/Korea/China	$57
Egypt, Nile Valley	Egypt/Israel	$25
Ganges Valley	Syndicate	$40
India, E. Coast	Syndicate	$20
US, E. Seaboard	Arab Family	$90
Los Angeles, 2nd spray	Arab Family	$50
China	Taiwan	$50

Placer pointed to the second line of the chart.

"When Detroit Motor Co. placed the offer to have Japan sprayed, it didn't cover our costs. So, we solicited the Koreans and Chinese, and they jumped at the chance to raise the offer. There's a lot of room for creative exploitation of international hostilities through bid sharing. Anyone can play, no amount too small. Of course, we didn't let Egypt know that Israel had added money to Egypt's request to spray Cairo. Egypt might have withdrawn the offer."

"Now that the Mexico City operation is finished, there are more jobs out there and a lot more small bidders not on the chart. The money's just lying around, fellas."

As Regis left the briefing, he wondered how to tell Colleen about all the new offers of money and targets. And he couldn't help but think about how much money his law office might make.

30

THE BALL GAME

Soon after the ag pilots sprayed Los Angeles, the Atlanta CDC epidemiologists identified pockets of sterilized males. The geographic edge of one of the pockets happened to follow a street that marked the edge of the territory of a large gang called the "Locos."

This produced conversations in the barrio about whether the Locos remained a potent force.

"Hey, man, what about this spray crap. You really think we can't knock up our women?"

Another laughed, "No problem. We just gotta try harder." No one else laughed.

"Hey, Jorge. You been trying to get your old lady pregnant for quite a while. Any luck?"

Jorge jumped on his questioner, hitting him and screaming curses.

"What's the matter with him," a gang member asked.

"Jorge's old lady moved in with a 'Blood' last week. She said Jorge had lost his huevos and she wanted a real ball bearing male. Jorge swears he'll find her and kill her."

The Atlanta CDC announcements of sterilizations throughout the gang areas led more and more of the Locos' women to defect to the nearby Bloods. The Locos grew desperate. The emergence of a new category among them disrupted status relations among the Locos. The new group of members were called "punks," previously low-status gang members who escaped the spray.

Some women associated more with these punks, because they carried live sperm. One infuriated ex-Locos leader attacked and cut off the punk's advantage. Sociologists at nearby Cal State LA talked of a time of great role conflict for the residents of the Barrio.

Amidst these troubled times, a fleet of 5 cut-down pastel convertibles entered the Barrio. Chicanos from another gang, doped-up and excited, filled the cars—proud studs who clutched their crotches and taunted the Locos. "Hey, look what I got, cuate. Your balls in my hand. And your bitches in my bed. Can any of you fight better than you fuck?"

The inflamed Locos poured into the street and engulfed the five cars. They pulled the riders from the cars, and stripped them of their clothes and genitals.

One week later, a group of Skin-Heads who aspired to become full-fledged Nazis, marched into the Locos territory. They carried signs claiming that since they were white they were more masculine than probably anybody. Again, the Locos moved down into the streets to disagree. The carnage continued until the Los Angeles Police placed the area off limits to general traffic.

Peace returned slowly when the women realized that if they took up with the sterile men, they could work and play longer without the baggage of children. At that point, the men began boasting of their "barrio vasectomy."

YOU CRITICIZE US?

Back in China, Chloris van Damme collected Colleen Jameson and Samuel Jonah Jonas and took them to the political indoctrination room ("education" room) the day after Ted and Susan left.

Damme and a translator talked intently for a few minutes before the meeting began.

Damme asked, "What is the meaning of the word you use to refer to the Americans? The word is 'Mob'?"

"It is a simple joke," Jonah said. "When the four of us negotiated with the American corporation, someone used that name."

Damme said, "This word is like Mafia or the Japanese Yakuza?"

"Those are organizations," Jonah said. "'Mob' is just an affectionate name among the four of us."

Damme and her colleague muttered balefully to each other.

"My dear American scientists. I would like to congratulate you on your accomplishments. You built the necessary production plant as a technology-transfer prototype." Damme struggled with these words.

Colleen and Jonah smiled politely.

Then Damme asked, "But, now, let me hear your assessment. How well do you feel you have done with the project so far?"

Colleen spoke when she saw that Jonah was not going to answer.

"Well, the technology of sterile systems is not well developed here. We adapted local equipment beyond what I expected to be possible."

"So you are not satisfied with our technological capability?" Damme asked.

"I didn't say that. I was pleased that the team could adapt your petroleum and nutrient technology systems to our needs," Colleen said.

"So, you also found our petroleum and nutrient technologies inadequate?"

"Please. You change the meaning of my words. You know that we adapted most of the equipment from other fields," Colleen said. "For example, we completely redesigned your vapor and fluid seals and gaskets. You know what I say is true."

"I know that I hear you complaining repeatedly about the industrial failings of the People's Republic of China. Now, you also accuse the PRC political emissaries of distrusting what you tell us?"

Colleen shook her head. "That is not what I said."

Damme raised her voice. "So now I am a liar?"

Jonah broke in. "Ms. van Damme."

"Mrs.," she shouted.

"Mrs. van Damme. What do you want of us? What is the next project? Do we continue here?"

Damme said, "With as much criticism as you have for the PRC and its representatives, why do you ask to continue working with us?"

"You want us to leave?" Jonah asked, startled.

Damme pounded her fist on the table. "Now you suggest that you should abandon your contract." She glowered at Jonah.

"Please tell us what you want us to do," Jonah said. "We are under your control. We will cooperate."

"I'm glad you have volunteered to leave," Damme said. "It makes my job easier."

Colleen asked, "What job?"

Damme said, "As you seem to wish, we will arrange your exit from the PRC."

"You are throwing us out?" Colleen asked.

Damme said, "We accept your request to leave."

"Where will you release us?" Jonah asked.

"Hong Kong is the closest port."

"How do we get through customs?"

"You should have thought of that before you launched your vicious criticism of the PRC," Damme said.

Jonah said angrily, "You attacked us!"

Now Colleen tried to placate her. "We meant no criticism. Could you arrange entry into the US so that we are not blocked by customs?"

Damme paused. "It may be possible. Is that what you want?"

Colleen said, "Yes. That and the rest of our fee."

Damme, leaped up. "Fee? Your fee? You criticize the way you are treated, break the terms of your contract, ask to be returned, and demand your fee?"

Damme faced them with a stern expression.

"Your fees have been paid. Be thankful that we paid you well beyond what you did for the PRC."

Colleen started to speak. Jonah squeezed her arm so hard it bruised her.

"Would you please send us to the US as soon as possible?"

Damme said, "Much better. Much better. You are an ungrateful pair. You are well paid. You criticize your benefactors. You ask for favors beyond the terms of the contract. We will be well rid of you."

Jonah asked, "When can we leave?"

"We will let you know when it is convenient," Damme said. "Report to the production facility, hereafter, until we notify you that it is time to leave."

As soon as Colleen got back to the production facility, she disabled an automatic high pressure relief valve system on the steam tanks. Then she wired a steam generator to overheat when she pressed a switch just before their departure. She thought the system would blow up about three days after they left.

Jonah begged her not to activate the switch.

"Shush, dear," she soothed him on what turned out to be their last night in China.

31

NOT SPRAYING JAPAN

"Mustn't underestimate San Francisco traffic," Regis Jameson told the stewardess. She hurried him to his first class seat next to Steve Placer and the aircraft began to taxi.

"Regis, sorry about all this hush-rush," Steve said. "I appreciate your filling in. How are things in China?"

"Well, I think they got the new plant working." Regis refused to use the word "virus" in public. "But, Jonah was contaminated."

"Ouch," Steve said. "You're probably wondering why I'm dragging you off to Australia. First we got Charlie Drummond's expense vouchers from Marseille. Now he's in Sidney."

At the airport, Drummond greeted them with wild eyes, and hurried them into his rented car. His square features and short, powerful frame were tense with excitement. "Hi, Steve. She's beautiful, fellas. Italian, small, compact, not finished, and nobody wants her. Hi, Regis."

"How long is she, Charlie?" Steve asked.

"One hundred eighty meters, fellas. That's 590 feet overall."

"Charlie, that's a ship, not a boat," Placer said.

"Well, yeah. But she's just a tiny thing. Most of them are twice as big."

"Big what, Charlie?" Regis asked.

"Aircraft carrier, fellas." Drummond beamed. "She's perfect."

"I thought we wanted a small mother ship," Placer said. "Maybe a tug or freighter."

"You still need some deck landing space, even for helicopters," Drummond said. "And helicopters require way too much maintenance for time flown and weight delivered." He was stuffed with technical information.

He parked the car near a seaport complex and hurried them out a long pier. There, twice as long as a football field, lay a ship, mottled gray and white.

"Needs paint, of course. One ramp up gets us to the below-deck area. Looks like a long empty building, huh." Charlie led them up another ramp, and they stood on the flight deck.

"Can you imagine the ag planes scootin' along, lifting off about there?" He pointed to an area well short of the bow.

"Where's the steering bridge?" Steve asked.

"That little thing there, just enough to sail it from Europe. She can travel under her own power, but the rest is a shell."

"Who built it?" Regis asked.

"The Italian government, 20 years ago. Called it the Giuseppe Garibaldi. Then the administration ran out of money. The Australians wanted it to replace the carrier Melbourne. So, they brought it here to Sydney and planned to outfit it with helicopters. But it has plenty of room for our little ag planes."

"How do you train farm pilots to land on this sea-going flight strip?" Regis asked.

"You just paint two lines down the deck the size of the ruts on a country road. Ag pilots have been landing on back roads for years." Drummond returned their incredulous stares. "It's true, fellas. We'll train 'em in light weather. They won't have to fly if the wind's blowing too hard."

"Can we lease this?" Placer asked.

"For a song," Drummond said. "The Australians intended to make it fully operational. But they screwed around, took a lot of time with renovation plans, and the Cold War ended. The government won't spend a dime on it, now."

"Would the government have any reservations about leasing the ship for spray activities?" Regis asked.

"Reservations?" Placer asked.

"Well, scruples, moral objections," Regis said.

"Moral objections?" Drummond laughed. "Look how they treated the aborigines. The kangaroos. Naw they don't have any scruples a'tall."

"It's not too expensive?" Regis asked skeptically.

Charlie looked annoyed. "Fellas, open up your minds a crack. Right now, she's just sitting rusting. Our friends in Royal British Shale said we can lease it with an Australian skeleton crew for the cost of maintenance and the men's contracts." Drummond then took them through the rest of the hull, showing them the plane storage area and the places they could bolt the crew's Portakabin.

Finally, he stopped, turned to them solemnly, and said, "If you fellas really don't like this one, there's others."

"The U.S. has a half-dozen this size, a tad heavier, but they are fitted out, and the Navy's proud of them."

"Have you looked at them?" Regis asked.

"Naw. I been trying to find a bargain. The French have a small one like this. The British have three."

"Slow down, Charlie," Placer said. "I'm still getting used to the idea."

"Ya gotta hear this last one. The Russians have three that are half-again longer. They'd almost give them to us to reduce their expenses. They can't afford ship maintenance since they went through divestiture and invited us capitalists in to help 'em. Of course, they'd be more expensive to run."

Drummond stopped and beamed at them. "What do you think?"

"My God, Charlie, you missed your calling," Placer said.

"Naw. I found it."

Regis caught Steve Placer away from Drummond.

"You sure you want an aircraft carrier? It'll be expensive, and fitting out takes time."

"It's an available technology," Steve said. "Don't underestimate Charlie Drummond and his country ways. With his banking connections, he could buy or borrow that ship. Also, we have friends in the maritime area around Japan who will help us. The Koreans still remember the Japanese occupation with great hostility."

"But naval vessels can sink a carrier," Regis said.

"No, no." Steve Placer laughed. "We'll arrange some business concessions, and we'll make it profitable to the top political echelon to ignore us."

"You can make that work?" Regis asked.

Placer laughed again. "It's the rule in conducting business overseas."

Regis shook his head. "Using a carrier seems a little grandiose. But we certainly could use it for spray projects anywhere in the world."

"How else can you deliver a fleet of spray planes off Tokyo and then Calcutta?" Placer asked.

When they told Drummond of their approval, he showed little emotion. "Good. We better quit screwing around and sign those contracts. That's too much money to wait long. I've already heard that French pharmaceutical companies are considering getting into the game. Right now, they are having the same arguments we are."

Steve Placer's boss at Detroit Motors, was jubilant about using the aircraft carrier to spray Japan.

"Be sure you include Japan's auto manufacturers in your targets.

Especially Honda. They think they're so good with that flexible management crap."

"Sir, I can't do that," Placer said. "We need the competition. Free enterprise makes America great."

"Don't pull my chain, Placer," his boss shouted. "Lot's of spray, right on Mitsubishi. Got it?"

"Yessir," Placer said.

A bit later, Steve Placer invited Charlie Drummond and Regis Jameson to Houston to review the plans for the attack on Japan.

"Our aircraft carrier arrived in California to pick up the ag planes. We'll call it the 'Shujutsu.' Means 'Operation,' as in surgery," Steve said.

"You're going to use the big spray planes? The old bombers?" Regis asked.

"That's right." Placer pulled down maps of Cheju-do, Korea, only 100 miles from Japan.

The new company, Kin Tama, Ltd, bought up big planes for conversion to water spray equipment and long-range service. When ready, the ship's cranes loaded them on the Shujutsu to go to Cheju-do.

The farm community at the wharf and the farmers who clustered around the foot of Cheju-do's tall volcanic plug thought the 1950 war had started again.

Steve Placer intended the name Kin Tama, Ltd to suggest a Japanese corporation as a cover for other Asian targets. Kin Tama, Ltd sought more contracts to "service" other nearby heavily populated areas.

The heavy aircraft of Kin Tama, Ltd sprayed from the lower end of the Japanese Islands north toward Sasebo, skipping Nagasaki and Hiroshima. Aboard the aircraft carrier Shujutsu, Charlie Drummond asked the skipper about their itinerary.

"We'll start up on Hokkaido and work our way south to overlap the earlier flights of the big planes."

The Japanese shipyard workers recognized the Shujutsu, since it put in for emergency repairs just before the project started.

Ferry pilots who worked the heavy planes back and forth from Cheju-do discussed each mission. Although top secret, the lowliest members of the flight teams payed close attention to each sortie.

"So where are we going?"
"To Japan."
"What for?"
"To disrupt the Japanese automobile competition with Detroit Motors."
"They got anti-aircraft guns?"
"No."
"Fighter planes?"
"Yeah, but the government grounded them."
"Why?"
"Business."
"Surface to air missiles?"
"Grounded."
"Business?"
"Business."
"Rifles? Handguns?"
"Illegal. A peace-loving people."
"My kind of mission."

When the first flights caused concern in Japan, the Prime Minister briefed political allies. "These attacks will sterilize millions of our male citizens."

The Social Democrats asked, "What will you do?"

"Shut down the media services during the attack," the Prime Minister said.

"You will conceal this attack?"

"That is correct," the Prime Minister said. "We will say that Asian airplanes illegally over-fly Japan to photograph our industrial areas, trying to determine why Japanese workers are so superior."

The Minister for Planning spoke. "This reduction in the birth rate will help with some of our food production problems. I advise that we take no action."

"We will not strike back?" a media adviser asked.

The Prime Minister grew indignant. "And start another war? The spray company is American. They warned us. They gave assurances . . . and money. If you mention the spray, you will embarrass your superiors. Is that what you want? We will blackout news broadcasts. The aerosol spray will not happen."

After the success of the first Los Angeles spray, the same Marxist Arab Prince found advertisers who claimed "Specialty crop spraying, no job too big or small."

When he made contacts and described the job, the spray pilots laughed at him. Then he mentioned the amounts of money he would pay, and they stopped laughing.

The Prince thought back over his experiences in American universities after he left the Los Angeles college which refused to change his grade of D to "the B". He recalled the slights and insults he felt at universities in Wisconsin and New York.

He offered to pay for successful sprays of the cities along the northern US boundary. Free-lancers could participate. Despite complex logistics, the contractors managed to lay at least some virus over St. Paul, Minneapolis, Duluth, Madison, Milwaukee, Chicago, Grand Rapids, Detroit/Windsor, Buffalo, Rochester, and Syracuse.

One problem occurred when a plane lost power over Chicago and called a "Mayday." Fortunately for the contractor, the pilot went into Lake Michigan and sank with his plane before anyone could reach him.

32

BABES IN THE WOODS

As good as her threat, People's Republic of China political agent Mrs. Chloris van Damme threw Colleen and Jonah out of the country. Every plane they boarded made them feel better. Finally, the Golden Gate Bridge led them to the bay and San Francisco International Airport.

Regis waited for them and spirited them away in his mobile home hunting van. He took them north to their parents' cabin near the Oregon border. Colleen and Jonah stormed with rage over their treatment in China and cried with joy and relief at being back and safe.

"Are we safe?" they asked Regis.

"I think you should keep to yourselves here in Klamath for awhile." He tossed them an envelope. "Here's enough money for a while, and the cabin's full of food. Don't go out much. Rest up."

They would have agreed to anything. Colleen said, "Call us. OK?"

"I have an excuse to come back to Crescent City," Regis said, "now that I'm the new Chair for the National Association of Riflemen/Wildness Club Coalition."

"You mean your crazy proposal for the Riflemen to join the Wildness Club went through?" Jonah asked.

"Amazing," Colleen said.

"Yeah," Regis said. "Now we're trying to join with the National Wildlife Federation."

Without thinking about it, Colleen called the Birnham Foundation from Klamath and left her forwarding address. She still hoped forlornly that her submission might win a prize. At Shevlin's request, the Birnham Foundation called Shevlin and gave him Colleen's forwarding address.

Shevlin immediately called the San Francisco Bishop of the Holy Father Church and gave him Colleen's new location.

"Have you heard any more news?" he asked the Bishop.

"Yes," the Bishop said. "I have to thank you. My superiors were quite concerned. They have prepared a program to help correct Ms. Jameson's moral lapses."

Shevlin asked, "Can you tell me about it?"

"I am not at liberty to say. Perhaps at a later date," the San Francisco Bishop said.

That night, Jonah especially savored being home. He held Colleen tightly and told her, "Let's forget about that money we lost. I'm happy when I'm with you, and we have more money than we need."

"Do you think the Chinese will leave us alone?"

"I think so," Jonah said. "They could have taken us out and shot us while we were there. Much cheaper."

Colleen shuddered.

VACCINATE A TREE WITH A SPIKE; SAVE IT FROM THE SAWMILL

In Klamath, Colleen and Jonah washed away the experiences of the China trip. They traveled up and down the coast, enjoying the scenery, trying to ignore the stumps and broken logs of the clearcut forests. They stopped along the river at a roadside bar and salmon smoker. They ate kipper smoked salmon, oily and pungent, and washed it down with beer. Out-of-work loggers drank and played pool. The bartender asked Jonah where they came from. He decided he had a message for them.

"Someone drove up from Berkeley this morning and asked about a couple like you two. You know anyone from Berkeley?" He looked at them with suspicion.

"I don't think so," Jonah said. "What did he look like?"

Colleen and Jonah listened in horror as the bartender described Jared Shevlin completely, from his crew-cut past his rings to his sandals.

"Did he say he'd be back?" Jonah asked.

"Yeah. He said he'd come back this afternoon and wait around."

Jonah told the bartender, "Don't tell him we're here. He's a Sierra Club recruiter." Colleen started to giggle until Jonah kicked her shin.

The bartender glared after them as Jonah dragged Colleen away. They rushed back to the cabin and printed two makeshift bumper stickers. After lunch, they returned to the bar and watched until Shevlin arrived. When he went inside the bar, Colleen drove behind Shevlin's car. Jonah jumped out and taped two bumper stickers on Shevlin's Caddie convertible. They read:

RUN OVER A LOGGER—SAVE A TREE, and,

SPIKE A TREE—STOP THE SAWMILLS

Later, several loggers arrived in a pickup truck with its own sticker:

SIERRA CLUB KISS MY AXE. The loggers read Shevlin's bumper stickers. One reached into the back of the pick-up truck and pulled a welding torch over to Shevlin's bumper. He lit it and quickly cut the bumper from the Caddie. The other loggers carried the bumper into the bar.

"Hey, Pete, lookee what we found. Who belongs to this?"

Jonah drove away when they heard the sound of breaking glass.

THE CONTRITE ONES

Many decades before New Mexico became a state, a group of poor Mexican dry farmers built scattered adobe shacks especially far back in the hills north from the trail crossing called Santa Fe. They named the settlement Sangre de Cristo, or, the Blood of Christ.

They were pious, intensely devout, and considered themselves members of the Holy Father Church. But their worship demanded more than most.

When a man learned he had been chosen for the Easter ceremony, he felt proud. He might have said to his woman:

"I will pull the cross to the church. I will mount it, and I will join our Savior."

"And if you die?" his woman might ask.

"I am a Contrito. I will go to heaven."

During the procession to express contrition, some whipped themselves, some carried stems of cactus on their backs, and the chosen one carried the cross. Thus they became "Los Contritos." At the peak of the ceremony, these peasants crucified the willing man whom the congregation chose.

Because of this rite, the priests of the Holy Father Church excommunicated the Contritos. Excommunication was emotionally vicious to the Contritos since it left their obligations to the church in place. It only removed them from access to redemption through the Holy Father Church.

Some who pass by Sangre de Cristo on Thursday night of Easter week report seeing candle-bearing processions in the hills near the darkened villages.

FISH CAUGHT

Colleen and Jonah began to feel much better after a few weeks among the streams and redwood trees. Colleen noticed the steelhead and salmon running upstream in the rivers, and she decided they should go fishing.

They bought their supplies in a tiny store with a hand-lettered sign above the fishing supplies which announced, "We don't give a damn how you do it in LA."

Colleen and Jonah tucked their hair into lumbermen's caps and pulled the bills down, just another couple of goddam lazy pot-smokin' tourist hippies. They took their spinning rods to Klamath's Terwer Riffle with its boat ramp and flight strip.

Colleen showed Jonah how to cast a spinner and try to keep it deep. He hooked the bottom on most throws, but slowly got better. Finally Colleen jerked her pole.

"Fish on!" She kept tension on the line.

"See if you can tell whether the line is moving," Jonah said. "I think you hooked a log."

They fished so intently they failed to hear an airplane land on the nearby river floodplain. Nor did they notice the advance of a black van toward the launch ramp where Colleen played the log.

Suddenly two large men descended upon them. "Folks, we need to see your fishing licenses and some identification."

"Fishing license," Colleen said. She peered straight ahead at the line where it entered the water.

"Yes Ma'am. Take her pole for her," he told the other warden. "And keep the tip up."

"It's moving upstream," Jonah whooped. "Don't lose it."

The warden took the pole from Colleen and followed the tip upstream.

The other warden spoke to Colleen and Jonah.

"It'll save you money, Ma'am, if you don't catch the fish. Now, step over here please, and show me your drivers license. Put it on the door jam there." As Colleen leaned over, the warden slid his hand between her legs, raised her feet off the ground and pitched her forward onto the floor of the van. He pulled his pistol, waved Jonah forward, and pushed him on top of Colleen. He snapped a handcuff on Colleen's wrist, pulled the other cuff under the seat brace, and snapped it onto Jonah's wrist.

He jumped into the driver's seat, kicked the starter, and bellowed, "Let's go."

The other warden trudged happily into view, carrying an eight pound steelhead. "Lookahere. Prob'ly got roe."

"Get in, dammit." They drove the van up onto the floodplain flight strip and quickly loaded Colleen and Jonah into the airplane. Fourteen hours, and two stops later, the pilot landed near Espanola, New Mexico. Another car would carry them the final leg of their trip to the village of Sangre de Cristo.

The Holy Father Church of Santa Fe reserved rooms at La Fonda, the oldest hotel in town, for a dozen prominent clerics, not named. One by one, they began to arrive from distant places.

33

TRIAL

The small flight strip had no radio operator present. A waiting car carried three black-suited young priests who untied Colleen and

Jonah one at a time and moved them to the car. At least these aren't ravening thugs, Colleen thought. Peaceful and polite, the priests ignored them.

"Where are you taking us?" Jonah asked.

"No espeak English," the priests said. They talked only to each other in Spanish, absorbed with, and commenting on, the strange arid landscape surrounding them. The road wound through low hills which looked green, but on closer inspection, consisted of loose, dry, sandy soil holding occasional pinon trees and chamisa bushes. The highway dipped through arroyos, dry stream beds that carried rare flash floods violent enough to tumble cars miles downstream. As they approached Sangre de Cristo, the barbed-wire fences along the road carried small wooden crosses.

The car turned off the highway and stopped near a small gathering of adobe houses, some with strings of red chile peppers hanging to dry outside the door. These villagers lived far up in the mountains and stared at this invasion of luxurious cars and stiffly dressed priests. Colleen and Jonah were placed in a hut near a larger morada, the village church. A guard stood outside.

Who kidnapped them this time? they wondered. They talked and dozed through a disoriented night. In the morning, the priests returned and escorted Colleen and Jonah into the morada, seating them in two handmade chairs facing an altar. A farmer came with them, smiled continuously, and occasionally gave them eye contact as he tied them loosely to the chairs. All ignored Colleen's and Jonah's questions. The farmer's old clothes and friendly manner stood in sharp contrast to the clerics'.

Left alone in the building, Jonah examined the plastered adobe wall, the small windows, wood floors and roof, and handmade tables and altar made of simple bare wood, polished where peoples' hands and bodies rubbed them. A carved Christ looked down, hands and feet conspicuously spiked to the cross, a spear-cut under one eye, and painted drops of blood running down his cheek.

Colleen focused on the sky beyond the window, and the fabrics and decorations on the altar, all hand sewn.

Three young men arrived, dressed in black with reversed collars, and seated themselves at tables near the altar, facing Colleen and Jonah, ready to take notes.

"Please answer our questions. You will have time to ask questions later and make statements in your behalf."

"Where were you born?"

"Tell us your parents' names."

"And the religious beliefs of your parents?"

"They baptized you?"

"Describe the first religious training you had."

"Have you ever had any religious insights?"

"Did your early school experiences include any effort to bring God into your lives?"

"How about other worship services with your parents?"

"Your friends?"

After an hour, their lives from birth through primary school lay bare.

"Did you masturbate as children?"

"Other sexual experiences?"

Colleen objected. The questioners continued, unperturbed, and wrote on sheets of thick paper.

A chime sounded, and a robed clergyman entered the room, attended by a young boy. He went to the altar and made offerings to the statue on the cross. After intonations in Latin, he turned to Colleen and Jonah and spoke with an Italian accent.

"Please be truthful. Lying will not help you. Your past behavior will be determined, so you must not lie. Give yourselves to the mercy of the Holy Father and reveal your sins."

He returned to Latin, with more imploring incantations. Then he and the boy departed.

Jonah asked the young priests who remained, "What did he mean, lying will not help us? Help us from what? Why are we being questioned?"

"You will hear a full statement of the sins you have committed at a later time. For now, you should lay a foundation for clearly explaining the grounds of your moral failures."

They ignored Jonah's further questions, and waited until he subsided. These questioners came from the outside world, cold and distant, unlike the sun-baked farmer from Sangre de Cristo who smiled as he tied them to the chairs.

"What do you recall of your early school training?"

"Did you have moral training at home?"

"Did you attend school continuously?"

"Where?"

"Were any of the schools religious schools?"

"The teachers were denominational, or secular?"

"Which denominations?"

"When did you have your first sexual experience?"

"With whom?"

"Tell about your feelings during the experience."

"After the experience."

"When did you begin to feel guilt at this first transgression?"

Colleen balked, and refused to answer. They threatened to place Colleen and Jonah in separate "interrogatories." Colleen decided she would rather keep Jonah present.

"Why do you call this a cleansing process?" Jonah asked.

"We give you an opportunity to review all the sins you committed in your life. It will make your later confessions easier, and will make your forgiveness rites more complete. Your spiritual salvation will be more deep and satisfying if you have provided full disclosures."

Jonah became frightened. That night he told Colleen he feared the questioning was part of a trial.

"But for what?" Colleen wondered. "Certainly we've done things for which we should be tried, but not for these 'moral transgressions.'"

"Right. Also, I note that I have slowly begun to feel more guilt-ridden. They are laying a trip on us," Jonah said.

"Yes, but I'm afraid to resist."

"Me too," Jonah said.

Midway through the second day, the inquisitors had trampled through the intimacies of Colleen's and Jonah's teen years. Each mid-morning and mid-afternoon, there was the interruption for Latin ritual.

It was during the afternoon of that second day that a much more imposing interruption occurred. A fully-costumed cleric entered with two attendants carrying chairs. The regular drab interrogators retired to the background. The scribe gave his notes to the newcomer. The big-bodied cleric spoke with a Spanish accent.

"I am Bishop Belen. I see that you refused to speak fully on sexual matters. Why is that?"

"My past sexual life is only my concern," Colleen said.

"Ah. Then you submit yourself to no other moral judgment than your own conscience."

"Uh, I guess so. Yes."

"And what if your memory or your conscience is faulty? What if you have occasional lapses? Is this possible?"

"I don't know. Perhaps."

"Then how shall we make judgments about your moral excellence without full knowledge of your past behaviors and your feelings about these behaviors?"

Colleen shook her head in dismay, and fell silent.

"And you, my son. It says you had a homosexual encounter at age eleven. Did you regret this?"

"It frightened me," Jonah said. "No one told me that this was common for young American males."

"In your studies, you both focused heavily on sex. Why is that?" Bishop Belen asked.

"We did not. We focused on the mechanism of biological change, and sexual reproduction is that process. Your church is excessively concerned with human sexuality with your prohibitions and fears," Colleen said, indignantly.

Jonah added, "Sex is a cornerstone of evolution. The organisms which reproduce most successfully replace those less effective. We study it because we are biologists."

The older cleric laughed. "Ah yes. Evolution. I sometimes forget."

He turned to the interrogators and tapped the notes with a fingernail. "You provided me with very little information regarding this significant topic." He raised his voice. "Will you please be more careful? This is most important to obtain a complete cleansing."

The interrogators whispered loudly, "Excellency, we will be more diligent. We regret this failure."

His Excellency abruptly departed, the attendants trailing after him. The scribes retrieved the notes from the floor where the offended cleric dropped them.

The interrogators made intensive use of the remaining hour of the day.

That night, Colleen said to Jonah, "You're right. This is a trial."

Jonah agreed.

"What do you imagine the punishment will be if we are found guilty?"

Jonah shuddered. He shook his head. They slept and dreamed of dreaded events.

On the third day, when Colleen and Jonah arrived in the church, five officials, backs turned to them, offered homage to the crucified image at the alter.

Finally the officials turned and settled into chairs with the assistance of attendants. The most elaborately garbed person spoke.

"I am the Bishop Albuquerque. I shall conduct the Ceremony of the Presentation of the Particulars."

By now, the proceedings became a tortuous confessional. A black depression settled upon Jonah.

"When you went to your secondary school, did you select what you would study?" Bishop Albuquerque did not look up at them as he spoke.

Neither answered.

"Come, come, now. You shall answer, my children." He pronounced it "cheel-da-ren."

He repeated the question.

Both Colleen and Jonah felt intimidated and frightened. They crouched in their chairs, unable to present their dissected selves for further examination.

"Now, now, this will not do." He waved his hand. His aides came and further bound Colleen and Jonah to the arms of the chairs. One of them applied an antique thumb screw to Colleen's hand, and twisted it down rapidly.

Jonah quickly shouted, "No, I'll talk to you. Please."

Colleen stiffened and groaned.

"Please. The question." Jonah searched for the question. "Did we choose our topics to study in high school? No. We did not. Our parents and teachers selected them."

Colleen moaned more loudly.

"Ah, so you selected your topics for the first time when you attended college?"

"Yes. Please don't hurt her anymore."

Bishop Albuquerque waved his hand. The aide removed the thumb screw.

Albuquerque asked, "Did you choose only the inconsequential courses such as crafts or physical education? Or did you choose the major course of study that you wanted?"

"We chose our major studies."

"And you Missy?"

Colleen had dropped her chin on her collar. Jonah leaned as far as he could and nudged her.

"Uh, yes. I chose biology as my major course of study."

Albuquerque asked, "You chose biology because you enjoyed it? Because you thought it important? Because it would earn you a living?"

Colleen said, "All."

"Of course. All of those. Did you ever think it would make you famous? A celebrity?"

"No. Never thought that . . ." Colleen said.

"How about your graduate studies? Did you try to stay with biology?"

Colleen said, "Yes. I applied for a scholarship."

"Did you want to keep doing the studies which the scholarship made possible?"

"Yes. The work held my interest," Colleen said.

"You liked physiology?"

"Yes."

"You wanted to study the physiology of sperm?" Albuquerque asked.

"Not exactly." Colleen appeared to equivocate. "The problems became reduced. They constituted a closed system, and could be studied as independent of other attached tissue."

"So of your own free will you chose to study the physiology of movement of the male sperm cell?"

"Colleen, I uh . . ." Jonah said as he began to get the drift of the questions.

Albuquerque turned to Jonah. "You will have a chance to speak, young man. Miss Jameson, you applied for the Birnham Foundation Prize of your own volition? In other words you wanted it, and wanted what it would provide for you?"

"Yes," Colleen said weakly.

"So you wanted to discover what you, in fact, did discover?" Albuquerque emphasized. "You wanted to be able to control the movement of the human male sperm cell?"

"No, I wanted to understand the physiology of the cell."

At the end of the afternoon, the questioner had examined Jonah in similar fashion. Jonah was led to agree that he wanted the Biosolz contract.

Far from the deserts of New Mexico, off the coast of Japan, Charlie Drummond stood as far forward as he could get on the bow of his new aircraft carrier, the Shujutsu. In the clear night sky, the horizon and stars rolled overhead from side to side. The bow moved like a giant corkscrew as the ship pitched and yawed its way from the southern tip of the Japanese archipelago toward Cheju-do. Drummond's thoughts whirled.

How could he go back to banking? He never felt so alive. The ship enthralled him, the physically challenging tasks and equipment to manage, the press of wind and waves. This must be what young men felt in times past when they first went to sea in ships. Finally, he went back to the bridge deck.

"How long does this feeling last?"

"What's that?" the skipper asked.

"The stars, the wind and waves, the ship's motion?"

The skipper smiled. "It lasted until my first big storm. A full gale, scared the hell out of me. But I know what you mean."

Drummond peered out the window in the dim red light. He watched the GPS digits slowly advance and felt himself, the boat, and the compass box swing together as a unit underneath the suspended needle.

"Where's our next job?" the Skipper asked.

"Up north a bit," Drummond said. "The Chinese want us to lay a swath over rice paddies. They want to see if the virus lives for a while in the water."

"Then where?" the Skipper asked.

"They also asked us to have a go at Taipei. God, you'd think the old resentments would eventually go away."

"Taiwan? That's next?"

Drummond said, "Maybe not. They have an air force."

"Yeah, but wouldn't they be risking their oil supply to attack us?" the Skipper asked. "Kin Tama, Ltd is owned by two oil corporations."

"That's right, if they are paying attention," Drummond said. "Are you ready to risk our baby on that?" He patted the bulkhead of the Shujutsu.

"By the way," the Skipper asked, "who is that throwing his guts over the stern?"

"Damn, I almost forgot. That's Flavin Taylor, one of our people that came along for some basic training," Drummond said. "I better go throw a line around him. A feller could fall overboard, back there."

Meanwhile, back in Berkeley, Professor Jared Shevlin's secretary handed him the telephone.

"Professor Shevlin, my name is Smith. I am told that you encountered the Jameson virus. I wonder if you can tell me anything about the mechanism of variation to other strains? My company is looking for consultants who are familiar with this problem."

"Certainly. I served as faculty advisor to the project which led to its discovery," Shevlin said. "What does your company do?"

"I would be pleased to tell you about my company. Is it possible that we could meet and talk? I think we have some ideas that could be financially profitable for you."

Shevlin invited them to his expensive home, high in the Berkeley hills, with its lovely view. When they arrived, he welcomed into his living room "Mr. Smith" and Mahmoud, the man who killed Michael Bearing.

34

VERDICT

Colleen examined a horned toad that chased a fly into their prison. The rust and beige colored disc-shaped lizard with leathery spines over its head and body surprised Jonah. Colleen picked it up and it secreted a drop of blood from the corner of its eye. "That usually frightens people," Colleen said.

"What does it eat?"

"It catches ants and insects with its tongue." She poked around the edge of the adobe wall and showed Jonah small yolk sacs, the animal's primitive eggs.

"I'd give anything to be out of this, and back to studying animals in the wild," Colleen said.

Jonah strained to get a view of the southern end of the Rocky Mountains, blue and jagged against the horizon. Pretty place to die, he thought.

Then they were thrust back under the sour scrutiny of the wrinkled accusers who pawed through their emotional lives.

"I am the Bishop de Magdalena. I serve God in Mexico City. I will complete the presentation of the Bill of Particulars." He spoke with a strong Spanish accent and watched their faces when he mentioned "Mexico City." Disappointed, he continued.

"Miss Jameson. Did you work especially hard in your early college years to win the scholarships awarded to you?"

"I didn't think about it. I enjoyed the coursework so much, I didn't think about scholarships."

"Ah, yes. A simple, happy child, playing joyfully in the meadows of academia." His voice carried little emotional shading. "When you obtained your Bachelor's degree, did you seek funds to continue?"

"I learned of a graduate scholarship at Berkeley and applied for it. They paid tuition and a small monthly grant for expenses. Out of state tuition was too high. I couldn't continue without it."

The Bishop of Magdalena called for water, wiped his brow, and continued.

"Tell us in your own words the events which occurred after you discovered the sperm virus."

Colleen thought back, and started slowly. "I sent the prize application to the Birnham Foundation. I thought this could give me the funds to complete my doctoral studies and continue my research."

She went on. "Then Jonah told me that he saw another student looking at my data and record books. Shortly afterward, someone killed that student in his home. This frightened me and I wished for a sheltered laboratory."

Colleen did not discuss the violence. She focused on her need for a laboratory like Biosolz to finish her work. China gave her another step toward independence so that she could continue her research.

Bishop de Magdalena asked, "The money was an important consideration, for research, of course?"

"Yes, we did want the money. We wanted the security. We wanted to study the virus. I still want to continue my studies. If you let me go this minute, I will immediately return to my studies."

The Bishop chuckled. "Indeed. You would like to continue? But you are here with us instead." He paused for a moment, his expression more serious.

"All right. Now would you please tell us about Israel and Baghdad. Then about Los Angeles. And Japan." His voice began to rise. "And when you are finished, you will especially tell me about my people in Mexico City!" His voice rose with rage as he pounded on the table and shouted "Mexico City."

Colleen and Jonah flinched. They sat mute.

"No? You refuse to tell us? You will tell us, now!"

Jonah spoke in barely audible tones. "We had no part in the decision to spray those places."

"Were you there when the decisions were made?"

"When I heard of it, I threatened to tell international authorities," Colleen said.

"And why did you not," the Bishop asked.

"The group convinced me I would be condemned as part of the group, regardless of my intentions."

"So you didn't try to stop your virus from being sprayed over all of those people." The Bishop sat back to watch the them react.

Colleen looked down at her hands in her lap.

Jonah asked, "How many did the spray expose?"

The Bishop de Magdalena said, "Five and one half million of God's children in Mexico. One hundred thousand in Los Angeles. Two hundred fifty thousand in Israel. Three million died from the bomb blast."

"The bomb blast?"

"Oh yes, the Israeli A-bomb dropped on Baghdad in retaliation for the spraying of the Israeli coast."

"Oh God," moaned Colleen. "We were in China."

"So you know God's name?" Bishop Magdalena continued.

"Finally, planes sprayed the Japanese islands, exposing an estimated 19 million people. About 25 million total."

Jonah said to Colleen, "Biosolz must have sold it all along. After we left for China, they increased their production rate."

The Bishop de Magdalena said, "Your American Central Bank recently purchased Biosolz' entire operation. You knew this?"

"No. Nothing," Jonah said. "When we arrived at Biosolz, our contract said we must isolate ourselves for the first months of our contract. The isolation was supposed to assure we would finish our work before the contract completion date. Then we went to China and heard little news."

"So you scarcely knew of these spray programs," the Bishop asked, leaning forward with an ugly face.

"Not Japan," Colleen said.

"Could these things have happened without you?" the Bishop asked.

"We only produced the virus," Jonah said.

"For money, you produced the virus. But without technicians such as yourself, there would have been no virus. Am I correct?"

Colleen shrugged. Jonah shook his head, no.

"Well of course you must agree." He paused and looked around.

"Do you not detect in this room, as I do, the stink of Nazis? Nuremberg? Genocidal irresponsibility? Without the two of you there would have been no spray attacks." The Bishop leaned back, turned aside, muttered an imprecation in Spanish, and dismissed them. He departed laboriously with his aides.

Colleen and Jonah sat shaking their heads.

GUILTY?

The next day, Colleen and Jonah sat in the church and watched with increasing fear as an even larger throng of clergymen assembled. Finally, three older, red-swathed officials entered and took the center chairs on a raised dais. The now familiar bells sounded. In the back, a dozen elders from the village entered and stood quietly.

After completing a prayer ceremony, the central figure spoke. "I am Arch Bishop MacKinstry of the Southwestern United States region. The Inquisitors will now present the evidence."

In rapid order, Colleen and Jonah's questioners reviewed the phases of questioning.

"Yes, they described their early childhood. All seemed innocent."

"Later in college, they became caught up with desire for increased prestige, money, and the acquisition of technical skills in human biology."

"They sought further recognition for the development of the virus. She applied for a prize."

"Who provided this prize?"

"An organization called the Birnham Foundation. We aren't sure of the source of their money."

"Proceed."

"Their next two jobs allowed them to earn ever more money to develop the virus, first for the American Biosolz Corporation, then for the Peoples' Republic of China. They disclaim responsibility for use of the spray. Their arguments seem self-serving, Your Excellency."

"Very good. Is there more?" Turning to Colleen and Jonah, he asked, "You wish to speak?"

Colleen shook her head, no. Jonah said, "No."

"Good. Is the Tribunal ready to give its judgment?" Turning from side to side, he nodded and said, "Bring the ballots."

Two aides lifted the handles of an ornate box. They approached the lesser functionaries on the ends of the dais first. Each reached out, opened a drawer in the box, selected a ball from the drawer, and dropped it into a hole in the top of the box. As the last cleric cast his ballot, the box was set on a stand. The bell sounded again.

Right out of the Middle Ages, Jonah thought.

Now, the center three clerics rose, opened the box lid, and gazed inside. Each returned to his chair.

His Holy Excellence rose again, made the sign of the cross, and asked, "Is the ballot clear or cloudy?"

"Cloudy," said the three who looked in the box.

"Thank you," said His Holy Excellence. He drew himself up.

"The unanimous judgment of this formally appointed Tribunal of the Holy Father Church, and ordained by the Holy Father himself, is that Colleen Jameson and Samuel Jonas, are guilty of the mortal sins of causing death for profit and self-glorification."

"No," Colleen burst out.

The clerics ignored her, and filed out, each pausing in front of the altar, with an obeisance for the dying figure nailed to the rude cross.

MOTHER SHIP

Soon after the Mexico City spray operation, Steve Placer decided they needed their own source of virus spray. He arranged for Detroit Motors board members to apply influence and pressure to Biosolz board members and carried out a hostile takeover of Biosolz. Mary Henly fought but lost her shares, her job, and her power. She had only her money and Sean to console her, and Sean soon left.

With the new crew installed on the Carrier Shujutsu, Kin Tama, Ltd. prepared to make a mobile virus production module for the aircraft carrier. Steve Placer oriented the new manager.

"We want you to build a virus production unit which we can unbolt and put on a few lowboy trailers."

"Length doesn't matter?"

"Not so much. We want to move it aboard the Shujutsu, and we have large spaces available."

The Aircraft Carrier Shujutsu was self-contained. Now it could continuously incubate the virus solution and deliver it as a spray anywhere in the world.

35

SENTENCE

Colleen and Jonah awoke the next morning to the sound of hammering and sawing outside the church. Jonah asked the guard what the carpenters were building.

"Crosses, señor."

"How many?"

"Two, señor," he said, leering slyly.

"What for?"

The guard sobered. "Today comes Papa, señor."

"Papa? He's traveling this far?" Jonah asked, not sure he heard correctly.

"Sí, señor, in your honor," laughed the guard.

Later, Jonah told Colleen, "This may be very serious. The guard said Papa is coming today."

"That's impossible. Coming here?" Colleen asked.

"That's what he said."

For the first time, the two of them could walk about for part of the morning. They approached the local farmers and asked their location.

"Sangre de Cristo."

"What state?" Jonah asked.

"New Mexico."

"Ah. The mountains are beautiful."

"Yes, it is very cold as you go up toward the peaks."

The locals seemed genuinely pleased to speak to them, but often crossed themselves, and bowed slightly as Colleen and Jonah passed by. Jonah tried one more question.

"What is the ceremony today, and why is Papa coming?"

"He comes to bless the ceremony of salvation and sanctification." The last words were spoken rapidly in Spanish.

When the guards took them to their usual chairs, they tied them only as a pretense. For the first time the cords remained loose enough to slip off, and the guard muttered apologies as he worked. He seemed to consider the restraints unnecessary.

The five resplendent clergymen returned from the day before, with a conspicuous difference. The middle chair waited empty, swathed in ornate white cloth.

Jonah counted perhaps thirty of the local citizens standing in the back, including some women.

The ceremonies began with the bell. All of the garbed clergy attended the altar with long invocations. Jonah heard the word "sanctification" and "salvation" repeated. His dread grew, and he clutched Colleen's hand.

"Looks pretty bad, huh?" she said.

"Yes it does. We'd better pay attention and be ready to move. By the way, I love you. I don't know how we got into this."

"I love you too," she said and squeezed his hand.

After more ceremonies, the senior clergyman addressed the throng.

"Heavenly Father, today we are assembled to perform the Rites of Sanctification and Salvation in the land of our ancient children, these contrite souls who have consecrated their lands for centuries, giving evidence of their love of the Heavenly Father and showing faith in God's mercy through their deep religious convictions."

Local eyes gleamed as they nodded agreement.

"I regret that the Contritos suffered excommunication from our Holy Father Church. This ceremony is to show support for the wisdom of your ancient practices of worship. We understand that there are three of you here, today, who have been sanctified?"

Three men hobbled forward with only limited use of their arms and legs. A young boy helped the frailest, most ancient one.

"We know of the strength of your convictions, and we commend you for it. You provided leadership and moral guidance to your community through the years."

Since they were sanctified? Jonah wondered.

"Please, tell us your names," the senior clergyman said.

One of the men took a step forward. He spoke with a low voice.

"I am Francisco, the Saint of Compassion," he said. "I care for those who have troubles of the spirit in our village."

The local throng nodded.

The second man spoke. "I am Felipe, the Saint of Truth. On the cross, I lost my ability to deceive. They know I only say what is true." He inclined his head toward the villagers.

Again the throng murmured agreement.

The last man said, "I am Miguel, Saint of Wisdom. Since my crucifixion, all listen to what I tell them. I am usually right."

"Correcto," a villager said.

"Thank you, most honorable Santos." Smiling broadly, His Excellency said, "Now, let us turn to the matter at hand."

He turned toward Colleen and Jonah.

"We summoned these two as a demonstration that the citizens of Sangre de Cristo have great faith. The Holy Father Church should not have turned away from your practice of crucifixion. We are here to acknowledge that your practices are wise, that we should return to your virtues."

The listeners from the village exclaimed with surprise. A noisy babble broke out. They never felt approval from anyone before, always outcasts in their religious rites.

Santo Francisco, the Saint of Compassion, spoke. "Please, could we hear more?" Silence suddenly fell.

His Excellency said, "We shall explain what has happened, then proceed with the ceremony."

"These two young people practiced what they call the science of biology. They discovered a disease which prevents people from having children." The crowd exclaimed.

"They let it be used by many people, in other countries, where people hate each other. This lost the Holy Father Church many souls.

"Did it lose as many souls as the war?" asked the Saint of Wisdom, a grizzled war veteran.

Ignoring him, His Excellency pointed to Colleen and Jonah. "These monsters defied God's will. God tells you to 'Go forth and multiply.'"

Muttering arose. A woman among the villagers said clearly, "We practice family planning in this town."

His Excellency seemed not to hear. "They have prevented the births of many millions of babies. For this evil they must be provided the ceremony of sanctification and salvation."

He turned to Colleen and Jonah.

"This Holy Tribunal found you guilty of breaking the moral laws of the Holy Father Church. Our Most Holy Father has come from the other side of the world to consecrate this ceremony."

The village residents exclaimed in awe.

His Excellency went on, "For your sins, you will be removed from this building and crucified at the moment these proceedings are concluded. Through this crucifixion, you will first be sanctified and freed from the afflictions of your sin. You will also receive salvation from eternal damnation."

His Excellency said, "Are you ready to confess, to receive the salvation of Holy Father Church, to be saved from guilt in the eyes of God?"

Jonah stood.

"No! I will accept nothing from you. Not from the church that excommunicated these honest people." He waved his arm toward the locals gathered in the back of the church. "Not from a church which called them unworthy to worship. Not from this Holy Father Church." Jonah pointed to the assembled clergymen.

Heads nodded in the throng.

His Excellency cried, "Enough of your heresies. Take them out and start the ceremony. Do it now," he demanded.

"Wait," the Saint of Truth said. He turned to Jonah and Colleen.

"Say more."
Colleen leaped up. "We studied the disease. We never wanted to spread the disease, and we hope to find a vaccine. Does this community find that sinful? I did not create the virus. I found it in God's world. We have done nothing immoral." She sat down and put her head in her arms. Women in the crowd murmured sympathetically.

The Saint of Wisdom turned to Jonah. "And you, please?" The clerics frowned and shook their heads.

Jonah spoke clearly and loudly to the villagers. "Your crucifixion ceremony has been part of your traditions for generations. But the Holy Father Church turned its back on you long ago."

Jonah continued. "If you put us on the cross, you will make us martyrs. Do you want us to become martyrs? We do not live with you. We are not of your religion. Will you benefit from our salvation?" The locals listened intently, hanging on Jonah's words.

"Which church needs the new vision, the new commitment? You people of Sangre de Cristo? Or the Holy Father Church? Who excommunicated you and now tells you that they should not have barred you from the church?"

"Ah." Jonah touched a nerve. A pulse throbbed in the group.

"Which of your churches is most successful? The Holy Father Church, with parents unable to care for their millions of unwanted children all over the world? Or your church, here with your loved and healthy children?

"Our church cares more," a villager said.

"Which church stood by to watch farmers pushed from their lands all over the Americas?"

One small village woman shouted from the rear, "Right on, brother."

Jonah continued. "If you want to help the Holy Father Church, give them a rebirth, a new vision. Provide the Holy Father Church and the whole world with new martyrs, those who are best known, those who are held in highest esteem." He paused.

"Give them these heroic volunteers." Jonah turned to the arrayed clergymen and swept them with his arm. They shrank from him.

Colleen looked outside and leaped to her feet. "Most especially, you should provide the ecstasy of martyrdom," she pointed, "to him!"

The door burst open and a resplendent Papa tottered in, blinking in the interior gloom, making the sign of the cross to the local worshipers.

Sunbeams followed him through the open door and focused upon Papa like a celestial spotlight. It transformed him and his white robes into a lustrous, radiant sight.

"Ahh . . . " breathed the throng, enthralled by the beatific vision.

36

CRUCIFIXION

"Flee, Your Worship. These ignorant heathens want to nail you to a cross," His Excellency shrieked.

"You call us ignorant heathens?" shouted the Saint of Truth. "We waited long enough since the excommunication of our ancestors. Today, we will reconnect ourselves with the highest authorities of the Holy Father Church."

"Bring all the crosses," shouted the Saint of Compassion. "Papa can provide absolution for all."

More people pressed into the church.

"We will show the Holy Father Church our new vision of the future. We plan our families. We only have children we can feed. You will see." Saint of Wisdom pounded his cane, shouting, "Let it begin!"

"Now," the three Saints raised their fists in unison. The local citizens, 200 strong, roared through the church, carrying their colorfully garbed, struggling visitors outside. The villagers brought a dozen more crosses, older, bloodstained, worn where dragged on the ground.

As the church emptied, Colleen and Jonah freed themselves, ran to a window, and eased themselves to the ground. They ran behind a shed. "Which way?" Jonah whispered.

"The highway is over there." She pointed.

"Come on. I'm not going to watch this." Jonah ran down the nearest ravine, away from the village.

Suddenly, a scream from the village turned to an awful wail. Jonah and Colleen stopped in horror. They heard a mallet thunk! punctuated by another scream. Blows and screams followed each other until Colleen covered her ears.

"Come on." They ran down the dry wash until they had to stop for breath. Jonah looked back.

"My God. Two crosses are up with someone hanging from them."

Suddenly Colleen's hands clamped on his arm. Colleen whispered, "Someone's following us. Come on." They came to a bridge and crawled behind the concrete pillars which supported the span at the bank. Someone ran onto the bridge while a teenager searched below, approaching where they hid.

Colleen and Jonah clung together trying to hide behind a pillar. They could hear a searcher breathing as he edged toward them. A car drove up and stopped on the bridge. A voice broke out inches from them. "Here they are, Santo."

The voice of the Saint of Compassion called. "Please come to the car. I must talk to you."

Colleen shook her head and refused to move.

"Please hurry. You are not safe here. I will take you to Santa Fe." Santo's urgent voice convinced Jonah. He started up the embankment.

As Jonah neared the car, Santo said, "Good. I must talk to you. Where is the lady?" Jonah looked at the old car with the empty back seat and a young driver and Santo in front. Santo glanced back at the village. "Please tell her to hurry."

Jonah ran down to Colleen and coaxed her out of the riverbed into the car.

"Vamos," Santo said. The driver pulled away down the dirt road toward the highway.

Santo asked, "Where will you go?"

Jonah answered, "California."

"We will take you to Albuquerque," Santo said. "You should leave this area."

"Thank you." Jonah said nothing more, afraid to disrupt the plan. Colleen became more peaceful, gazing out at the country familiar to her youth.

When the car approached the outskirts of Santa Fe, Santo stopped at a gas station. Jonah still had his wallet from Klamath and paid for a tank of gas. Colleen telephoned Regis in San Francisco and asked him to arrange for airline tickets.

As they drove, Santo tried to turn his scarred body, gave up, and adjusted the rear view mirror so he could see them. "I ask you to tell no one about today," he said. "My people seldom speak of the Holy Father Church. I myself have no more belief."

"You didn't expect what happened?" Jonah said.

"Oh no," Santo said. "Priests are not welcome in Sangre de Cristo, but I did not expect this." He shook his head. "All that ceremony, the prayers, the costumes, the hatred and accusations awoke old, deep resentments in us." Santo became less agitated.

"We live peacefully in the village. Our midwives train as nurses and keep a clinic. A few young people leave Sangre de Cristo, and no one seems to return. But mostly we are a stable community."

The car rattled down an endless steep grade. Santo twisted himself to look out. "La Bajada." He and the driver pointed at the dramatic landscape. Flat mesas stood all around, their sides near vertical, with an occasional barren volcanic cone. The rock colors showed continuous variations, greens to blues and rusty reds, to beige and white. Geologists carried color charts to describe them. Color worshiping painters saw the desert, and, stunned, stopped to struggle with their oils and watercolors, some never to escape. Santa Fe became an art colony.

As they neared Albuquerque, Colleen spoke.

"Thank you, Santo, for saving us."

He smiled sadly.

At the airport, Colleen hugged Santo goodbye. Jonah thrust the money in his wallet upon the driver.

"Anda. Go, go," Santo told them.

Colleen and Jonah boarded the plane, into the unfathomable luxury of padded chairs, smiling faces, and hot coffee. After four days in the old church building, Colleen still felt the hard chair on her buttocks. Her head whirled.

On the trip to San Francisco, Colleen sat stiffly silent, deep in thought. After a time she burst out: "Jonah, I won't run anymore!" Passengers turned and looked at her.

Colleen tried to remember the research jobs she coveted. But images of the villagers surging into the church thrust all other memories aside. Disoriented, she shivered and clutched the arms of her seat.

"Are you all right," Jonah asked.

"No, I'm not." She thought of the people who pursued them, threatened them. Why should she have to run from anyone?

"We haven't done anything wrong," she told Jonah. "We did our graduate work, and did it well. And they called us mercenary, and accused us of self-aggrandizement, abortion, infanticide, genocide, immorality."

She lapsed into her thoughts. How dare anyone say her

academic training produced immoral behavior. She learned to care about the planet and all of the life forms on it, every one of them, and she worked to protect them. Other people had better learn soon about the dangers of letting the population explode, she thought.

"What do you want to do now?" she asked Jonah.

"Nothing we've talked about, yet," Jonah said. "I'm angry at all of the people who controlled us."

"I wanted to do research." She turned toward Jonah. "They say we threatened human welfare? They say we wanted to reduce population? Why don't we really show them how to reduce population?" Her eyes glittered as she tried to quiet her rage.

"I like that," Jonah said. "We have enough money to buy all the peanut butter we need. Let's work on our own agenda, do exactly what we want."

Colleen reached over and kissed Jonah. They made a pact to pursue their own goals. Colleen thought more about the future. Threaten her and her natural world, would they? She had an image of early humans living as cave dwellers, afflicted by all the microbes and larger predators which killed them and kept their numbers down. Lacking those predators, humans must be their own, she decided.

Once more, Regis met them at the San Francisco airport. Colleen and Jonah looked years older. Regis saw them casting fearful glances about them.

"Are you still worried about people chasing you?" he asked.

"I am," Jonah said.

"People can buy the virus on the open market, now," Regis said. "They should leave you alone."

"Good." Jonah nodded, trying to feel reassured.

"Where have you been?" Regis asked.

"Maybe Jonah can talk about it," Colleen said. "I can't."

"No need. But there is a phone call I promised you would return." Regis punched his cell phone and handed it to Colleen. "Just tell them your name."

Wary, she put the phone to her ear. In a moment she heard, "Birnham Foundation, Tyler speaking."

"This is Colleen Jameson, I, uh . . ."

"Ah Ms. Jameson, thank you for returning my call. This is Moses Tyler speaking, I am President of the Birnham Foundation. I have some news for you."

"Yes?" Colleen choked.

"I just received a call from Dr. Jonathan Remmer. He is the Chair of the Birnham Prize Committee. Uh, are you there, Ms. Jameson?"

"Yes sir, I am."

"Good. I am pleased to tell you that the Prize Committee voted last night to award you the first Birnham Prize for your report of what is now called the Jameson virus. The Atlanta Center for Disease Containment has corroborated your initial description and predictions. Thus, you are credited for the discovery of a heretofore unknown virus. I am sure this pleases you?"

Colleen croaked,"Yes. It does. Very much, sir."

"Indeed. Well, we are pleased with the Prize Committee's decision and will issue a news release three weeks from today. We would appreciate it if you would not let the news get to the media before that date."

"Certainly, Dr. Tyler."

"Well, that is all I have to say now. We'll get all this to you in writing shortly, so we'll be talking in the near future," Tyler said. "Congratulations."

"Thank you, sir, and goodbye." She put down the phone and threw herself in Jonah's arms.

"What's the matter with this insane asylum?" she wailed. "One day we're being executed for religious crimes, and the next day we're heroes. You're in this too, you know," she told Jonah.

"What happened?" he asked.

"I won the Birnham Prize." She started crying.

"That's wonderful," Jonah said. "You'll be so important we won't get to argue with you any more."

"That's right," she sniffled, "No arguments."

37

ABOUT THAT DOCTORATE

Regis put them up in his apartment while they bought clothes and a car.

"While you were gone, I asked a private investigator how

anyone found you in Klamath," Regis said. "Apparently the Birnham Foundation gave out your forwarding addresses."

"That's right. I called them as soon as I got back from China," Colleen said. "I didn't think about it."

"From now on, use my office address," Regis said.

"Your partners won't mind?" Jonah asked.

"Mind?" Regis laughed. "The name 'Jameson' on the letterhead is drawing all kinds of legal inquiries to us. Virus related law is big business now. We've added staff."

"How would your company like to invest in an Institute?" Colleen felt exploratory and wasn't sure where she was leading.

"What kind of an Institute?" Regis asked.

"Oh, educational, with a planning and environmental policy legal focus." Colleen ad-libbed. "We'll do authoritative studies: the role of the autonomous citizen, the armed citizen who is a member of the NAR, in a society freeing itself from corporate control and industrial pollution?" Colleen shamelessly played to Regis' interests.

"You may think you're teasing me, but we're getting a different kind of client now, people who need protection from the corporate establishment. Write up something for me."

Jonah examined Colleen's face to see whether the prize had suddenly made her smarter.

"By the way," Regis said as he handed them a bundle of papers. "I found this stuff on the boat after you left. Michael Bearing's notes and his journal were in with your papers."

Jonah and Colleen voraciously read the entire pile, becoming angrier with each page. A letter from Shevlin asked Michael Bearing to find out what Colleen was working on. Bearing's notes described her lab work. Shevlin answered these notes and asked Bearing to send whatever else he learned.

Colleen snatched the phone and called Shevlin's office at the University. He answered.

"This is Colleen Jameson. I want to talk to you."

She heard him gasp. "Well how nice to hear from you. I wondered what happened to you."

"Meet us tonight at the Top of the Mark. Eight p.m."

"I beg your pardon. I have another engagement," Shevlin said.

"Break it. Jonah and I want to talk to you about the notes you wrote to Michael Bearing, just before he died."

"I don't know what you are talking about."

"Dr. Shevlin, be there, or you will have police in your office tomorrow morning."

"Well, I suppose I could cancel . . ."

"Eight p.m. tonight," she said, as she cut him off.

Jonah noticed she was much more direct, these days. She asked fewer rhetorical questions, like, Why are we being lied to, treated unfairly, and so forth? By now, she probably thinks she knows.

Ever the prima donna, Shevlin came late, even under duress. As he sat down, he said, "I've eaten. I'm in a hurry."

"You'd better have a drink," Colleen said.

Jonah leaned toward him and examined his face. "I don't recall those scars on your cheek."

"I was mugged," he said, irritated.

"Here in the city?" Jonah waited for an answer.

"No. Up north,"

"Klamath? The Riverside Bar?" Colleen asked.

"You set those thugs on me? I'll call the police."

"Worse may happen. Listen closely," Colleen said.

"Dr. Shevlin, you have not spoken the truth to your students, and that is a cardinal academic sin." She handed Shevlin copies of the papers in which he asked Bearing to steal Colleen's data and discussed their own application for the Birnham Prize.

Shevlin immediately threw them on the table.

"They prove nothing."

Colleen continued. "We want you to help us. If you refuse we will go first to the police and tell them of your possible involvement in causing Bearing's death. Then we will go to the University Ethics Committee with these notes to report your theft of student data. After that, we will file letters of complaint with the University Academic Senate, the Faculty Union, the UC Student Government, the University Administration, the Board of Trustees, the California legislature, the American Association of University Professors, the US Department of Health, and the American Virology Association." She paused for breath. "You will lose your tenure, your job, your employability, your professional reputation, and your relationship with those people you now believe to be your friends. It won't hurt your reputation with the students. They already know you." Shevlin's face, pale when Colleen started, turned ashen as she finished.

"You wouldn't," he gasped.

Colleen leaned forward. "I promise you we will. Do you still have that little notebook? Take it out and start writing."

Shevlin stared for a moment, then reached inside his jacket.

Colleen spoke slowly. "We are now giving you the chance to

help us as a good doctoral chairman should do, rather than harass and threaten us as you did." She paused for emphasis.

"In case you have forgotten, Jonah and I both finished all of our course work. We both got honors on our qualifying exams. You accepted us as doctoral candidates and approved our dissertation topics. You served as Chair for both of us. We nearly completed our dissertations. The committees only needed to approve our dissertations, and we would receive our doctorates. Am I going too fast?"

Shevlin wrote in his notebook.

"That is how far we got before you told my fellow students in your seminar that my work was not scientifically adequate. You did this while you stole my data, findings which have now won me the Birnham Prize."

Shevlin's head jerked up, terror in his eyes. He looked stricken.

"Ah. You didn't know that yet, but I did win the prize, you bastard," Colleen said.

"Jonah and I will finish writing our dissertations shortly. We ask you to suggest any changes required to bring our draft copies up to the standards required of your previous students. If any committee member wants us to make changes, we will be pleased to do so. When the dissertations are acceptable, we will meet whenever the committee wishes for the final oral exam.

"We ask you to make the arrangements, call the faculty members, schedule the meetings, and make sure the forms are filled out properly with your signature. This is busy work for you, but we'll be out of town. Did you write all that down?" Colleen watched Shevlin.

He raised his head and stared at her.

Colleen continued. "You can go now. Keep your secretary posted so she can give us progress reports. I'd rather deal with her than with you."

Jared Shevlin, PhD, MD stood, folded his napkin, and walked out of the room.

Colleen and Jonah then relaxed and enjoyed their first expensive dinner date. The wine tasted funny, but they liked the food.

"I'm getting insights I never had before, and I don't like them." Colleen said. "Things people say have double meanings. I used to accept the ordinary meaning of words. Now, I'm trying to feel the emotions behind the words when someone talks to me."

"Why not?" Jonah asked. "Word meanings differ because peoples' backgrounds differ."

"Well, my world is getting simpler," Colleen said. "There's

good guys and bad guys, and most people don't tell the truth." She paused. "I want to ask you something, but I'm scared."

"Then I forbid you to ask," Jonah said.

"Have you ever been upset when I get top billing? You know, more money? Now the prize?"

"No," Jonah said, "not if I can have part of the action. I can't tell you how much I enjoy our technical work together. Besides . . ."

Colleen shrank. Oh-oh, here it comes.

"Whenever you're in the spotlight, you speak with great clarity. People like to listen to you. So I look at this lovely, admired woman and I think, 'When we go home, we'll get into the same bed.'" Jonah took her hand and smiled at her. "That's exciting."

"Let's go home," Colleen said.

PART II

COLLEEN JAMESON, PhD: EXTREMIST

38

JAMESON POPULATION INSTITUTE

Regis Jameson invited Colleen and Jonah to talk to his two law partners. Compared to Regis' loud clothes and plain-spoken manner, Colleen saw the partners as colorless business types, with bland, hard-to-read smiles. Regis claimed they cultivated a shy, deferential image for the courtroom. More deception, Colleen thought.

"But why would your law office put money into a Population Control Institute?" Colleen asked.

"For the same reason corporations set up the American Enterprise and Heritage Institute," Regis said. "To influence the public using tax-free money. We'll keep our name before the public, develop a favorable social image, and legitimatize our efforts.

"You'd call it the Jameson Population Institute?" Colleen asked.

"If you agree," one of the partners said. "Tell us what you hope to accomplish?"

"All right," Jonah said, breaking in.

Ordinarily, Colleen spoke for herself. She resented being ignored. However, the day before, Colleen asked Jonah to take the lead in future discussions. She poured out a stream of feelings new to her. She didn't trust anyone anymore. She wanted to watch and listen to discussions, concentrate on non-verbal cues, and to speculate on people's motivations and feelings. She reminded Jonah that, "Mary Henly spoke sweetly and then battered us with management tactics."

So, Colleen asked Jonah to start the discussions for a while. She could always break in. Jonah agreed.

"We hope to combine information from the biological and behavioral sciences to ease the population crisis," Jonah said. "Our way of life and biological world are deeply threatened, and people don't know it. We need access to the media to tell the public. We need to show how environmental damage occurs when populations overshoot. We need to calculate the human carrying capacity of various regions under different standards of living."

"Would you temporarily take on specialists?" the partner asked.

"Yes, as adjunct and joint faculty appointments," Jonah said.

"How about students? Would you have training programs? Internships?"

"Yes, and those could be the strongest features," Jonah said. "We want to affiliate with a school like Arcadia State University in Northern California."

Regis handed out papers of incorporation for the Jameson Population Institute, devoted to planning, research, and development for population control.

Regis' partners agreed to visit the President of the California Board of State Universities. They would ask him to urge the President of Arcadia State University to affiliate with the new Jameson Population Institute, pointing out the research opportunities. Later, Jonah and Colleen would meet with the faculty decision leaders.

So that's how you squeeze college presidents, Colleen thought.

By design, Colleen, Jonah, and Regis arrived at California State University at Arcadia soon after the announcement that Colleen won the Birnham Prize.

The University President briefed the department chairs on the importance of the proposed Institute. Chairpersons might obtain joint appointments with the Institute. New faculty and students might be attracted. The Institute could accept commissions for studies. They would design new courses. Images of research monies and study grants danced in chairpersons' heads. Promotions might follow.

The President told the faculty that a determined and academically oriented small-town forestry and fisheries college might improve its research image. Colleen and Jonah marveled at the many incentives the administrators offered.

Discussions with department and division chairs focused on consulting and interdisciplinary research. Jonah noticed that no one asked about opportunities for students.

All a-twitter, the Biology faculty members demanded their own meeting to meet Colleen. Favorite students sneaked in.

Jonah told them about the Jameson Population Institute. Finally, the group got to its real agenda. A young woman spoke to Colleen.

"How did you discover the Jameson virus?"

"The Fred virus," Colleen laughed. She told them about her friend who couldn't become pregnant, and the almost accidental saving of the husband's semen specimen. She briefly mentioned the Biosolz and China experience. Then Colleen turned to the future.

"The Institute will focus on reducing population growth, not on family planning, not economic development, not even 'sustainable' development nor birth control, but upon reducing the birth rate."

"We are victims of overshoot. Our population grows faster than we can feed them. The world's long-term carrying capacity is under one and one-half billion people if the inhabitants of earth are to have a minimum standard of living. The human race may die out if we don't solve our growth problem soon."

A faculty member spoke. "But surely you don't approve of forced sterilization using your virus?"

Colleen remembered her decision to quit running and turned directly to her critic.

"What I really disapprove of is the starvation of those 50,000 children who died elsewhere in this world since you and I ate a hearty breakfast yesterday morning." A hush fell as Colleen's voice rose.

"Another 50,000 will die by tomorrow morning and every day thereafter until you and I find ways to reduce population growth."

"I also disapprove of the failure of our species and the best minds in it to self-limit our numbers." Colleen leaned toward her critic.

"But what really offends me are humanists who would allow overpopulation to destroy our ecosystem, our very world, in the name of morality and ethics." Colleen continued when the faculty member decided not to respond.

"The Institute invites you all to work on some aspect of population control. Saving the environment by reducing overpopulation must become the new college general education requirement. All students need environmental education. No subject matter discipline may ignore the threat to our ecosystem."

"We ask you to begin planning all aspects of the Institute operation. We'll want your ideas during the next few months." Colleen's pace slowed.

"Many thought the population problem insoluble. Now we know that it is only unpopular: the national media make it so. Help us find less punitive plans that will work. You may take and teach classes, do independent study, work on interdisciplinary teams searching for new approaches. We look forward to hearing from you." Colleen turned to Jonah, ready to leave.

"We'll keep you posted through your campus paper," Jonah said. As they left the group applauded.

Arcadia University jumped at the opportunity to affiliate with the Jameson Population Institute. Colleen and Jonah left San Francisco and moved into an older home near the Arcadia campus. They cobbled together a computer system and soon had rough drafts of their dissertation manuscripts. They made suggested changes, and sent them to Shevlin.

After a week of waiting in agony, a courier service brought their manuscripts with Shevlin's marginal notes. His suggestions were clear, specific, helpful, and straightforward. Jonah mused over his final draft late one night, thinking about Shevlin's changes.

"The son-of-a-bitch is really good, isn't he."

"Yeah, he is," Colleen said. "After all he put me through, his suggestions are objective. Each change clarified what I wrote." They submitted the new drafts to their committee members.

Jonah and Colleen received their degrees a week apart. Jonah happened to get his first and demanded that Colleen call him "Dr. Jonas" for a whole day.

Meanwhile, business boomed for Kin Tama, Ltd. The aircraft carrier Shujutsu continuously sailed on spray operations. Steve Placer told Regis, "Take a trip with us. Jonah's a sailor too. You'd both love it."

"I know I would," Regis said. "I'd love it so much I'd have a hard time going back to the office."

"We have a four-place plane aboard for observation," Placer said. "I'll fly you off the deck. Only chance you'll ever get to do that."

Regis realized just how much he would like to fly off a carrier and watch the ship fall away behind. "I have to get off the phone. I'm weakening."

Steve Placer laughed. "OK. Maybe next trip."

Colleen watched Jonah's face while Regis told them about the invitation.

"Don't pretend to me you don't want to go," she told Jonah. "Both of you are no better than soldiers of fortune. You'd do anything to play war games." Regis and Jonah protested.

The carrier Shujutsu had been at sea for several weeks. Flavin Taylor, Steve Placer, and Charlie Drummond felt like old salts. Though Drummond grew up on a farm his squat figure rolled from side to side taking the ship's motion in stride as if he were a life-long sailor. He squinted against the sun and wind with a perpetual smile.

Steve Placer looked like a movie star playing a carrier officer, with a bill cap that helped him follow planes close to the sun. He was slender and agile as he moved quickly around the deck. He loved to fly and brought a short-take-off plane aboard the carrier. He flew at all opportunities and sometimes accompanied the ag planes on spray missions.

Flavin Taylor was changed beyond recognition. The city dandy and elegant attire gone, he had a two month's length of hair in a pony tail. He carried a seaman's knife on his belt and a pair of vise pliers in his back pocket, now a competent seaman. The changes occurred on a day Drummond was on the bridge. Late in the afternoon, Drummond watched Flavin Taylor throw up over the rail. The ship pitched and rolled, with sheets of spray wetting down the stern, At least Taylor had learned to stand on the downwind side.

Suddenly, Taylor disappeared. Drummond focused binoculars on the stern. Pulled taut, the shackled safety line led over the rail. Drummond hit the speaker switch. "Man overboard! Deck crew aft."

Taylor's body hung like a wet rag, folded around the line to his waist. Once aboard, they carried him to the clinic and the first aid corpsman. Taylor revived, drank fluids, and regained his color.

"Say chaps," he said, "good of you to retrieve me. I kept banging against the hull."

For the first time, Charlie Drummond felt guilty for harassing Taylor. Taylor lifted a trembling arm and shook Drummond's hand, smiling at him gratefully. Taylor slept 10 hours, got up and put on a pair of jeans, a wind-proof jacket and sneakers, and tied his hair back. He ate in the ship's mess as if the day before had never happened. He never threw up his food again. He took turns on each work detail and quickly learned the different jobs. Like Drummond earlier, his sensitivities to the marine world became aroused.

One evening, they gathered on the bridge deck of the ship that Drummond now thought of as his personal yacht. They cruised northeast, approaching Taiwan, enroute to the home base off Korea. Drummond wanted to go home. They discussed the last mission, to spray dense population areas of Taiwan.

"How much is this job worth?" Taylor asked.

"Thirty million," Placer said.

"Are there problems?" Taylor asked.

"Well, for one thing, the Taiwanese Air Force has air-to-ground missiles," Drummond said.

"I'm ready, if you want to go for it," Taylor said.

"I'd hate to lose this ship," Placer said. "We've already doubled our profit margin."

Charlie Drummond turned and looked at both of them, grinning. "What the hell. Let's go home."

They immediately agreed.

Something held Drummond back from this last contract, probably the Taiwanese fighter planes. "Drinks are on me. Meet you in the bar?"

Flavin Taylor said, "I'll be right there." He wanted a last few minutes on the bridge on this brilliant night. The stars captured him too. He liked to watch the radar screen with a few fishing boats working around the southern tip of Taiwan.

His mind went back to the last six weeks off the mouth of the Ganges. They sprayed Calcutta and the flood-plain of the intertwined river channels as far as they could reach with the ag planes.

The inhabitants waved at them casually, reminding Taylor of the feelings of guilt he sometimes felt over casually infecting them. But he knew that since the Indira Gandhi government threatened them with sterilization, it was a common topic and frequent event in their communities and startled no one.

For compelling reasons, women preferred the laparoscope, with its small incision to isolate her ovaries, over vasectomy for her husband. If she had tubal ligation, and it failed, she could blame the surgeon. However, if her husband's vasectomy failed and she got pregnant, he and the community might blame her for being unfaithful.

"What's going on?" the helmsman asked. "Two boats are converging just off our port bow. See 'em? They must be making a transfer. What the hell?" He turned to the intercom.

"Skipper would you come to the bridge?"

Taylor watched the Skipper record the bearing of the two radar dots, now beginning to diverge. One of the dots turned back toward the island of Taiwan. The other headed across the path of the Shujutsu.

The Skipper immediately sounded a series of blasts from the ship's horn, and spoke into the radio. Safety and etiquette required an immediate change of course by the smaller vessel which could still maneuver to avoid hitting the ship. The larger vessel couldn't turn or stop rapidly to avoid a collision.

Drummond and Placer burst into the cabin. "What's up?"

The Skipper squinted at the radar screen. "A goddam small boat is on a collision course with us. What's funny is that it just went up to another boat which is now heading back to the island." He blasted the ship's horns again.

He rechecked the boat's bearing. "No change. He's close to us on a collision course." The Skipper throttled back, and put the helm hard over to port. "We can't stop it, but we can postpone it a bit. Maybe he'll slow down." To the helmsman he said, "Get a boat ready to pick up survivors."

Drummond, Taylor, and Placer followed the helmsman to a forward lifeboat davit. A searchlight beam showed an old fifty foot fishing boat, slogging along at a comfortable eight knots. Neither cabin nor deck showed signs of life.

Now only two hundred yards off their port side, the point at which the boat would impact the carrier slowly moved aft as the ship's rudder pushed the stern away from the fishing boat.

Charlie Drummond suddenly said, "Oh God." He moved to an intercom, shouting at the Skipper. "Her course hasn't changed since she departed the other boat? Right?"

"Right," the Skipper shouted.

"She's on auto-pilot. Nobody's on board. They dropped off onto the other boat. Could the impact sink us?"

"No," the Skipper shouted. "But get everyone back to the starboard side, just in case they packed that tub with explosives."

The boat approached within seconds of impact as the group turned and ran toward the starboard side. Charlie Drummond puffed, "I guess we should have told them we weren't going to spray Taiwan." As he ran he realized the fishing boat could easily carry more explosives than several wartime torpedoes.

A huge blast and sheet of flame came at them from the stern quarter. The explosion opened a large hole in the hull, blew out a flotation bulkhead, ruptured tanks, and ignited fuel.

The Skipper kept the ship moving forward to leave the flames behind, but the stern sank rapidly. The crew freed rafts and life boats. Everyone tried to get into the water.

The Shujutsu sank in a few minutes, its stern structures burning. A rescue vessel arrived and took sailors from the life boats. They found Flavin Taylor holding Charlie Drummond unconscious alongside a life raft. Steve Placer, the Skipper, and three crew members were never found.

39

INSTITUTE GOALS

Colleen and Jonah finally finished their doctoral work and turned all attention to the new Population Institute in Arcadia. Since this was Regis' old fishing area, he gave them a tour.

"Is it always gloomy like this," Jonah asked.

"Winter depressed everyone," Regis said. "When the loggers got laid off for the rainy season, they went on unemployment, drank, shot pool, and fooled around with each other's wives. Lots of divorces in the winter." Regis went on.

"After they cut most of the trees, the loggers disappeared, complaining that environmentalists cost them their jobs."

They drove to the new marina where the salmon and Dungeness crab boats sulked at their docks in Humboldt Bay.

"Is fishing off, too?" Jonah asked.

"The number of fish caught is way down," Regis said. "I used to ask the salmon fishermen if they weren't taking too many fish. They didn't think so. Some blamed the Indians even though the fishing boats brought in two-thirds of all fish caught."

After Colleen and Jonah moved to Arcadia, they tried to write the objectives for the new institute. To plan projects they needed to know the effect of the virus sprays. The United Nations World Health Organization provided estimates:

Area Sprayed	Males sterilized
Los Angeles	100,000
Ralph Nader U	4,000
Mexico City	4,500,000
Israel	1,000,000
Baghdad	2,500,000
Japan	20,000,000
Eastern US	4,000,000
India	70,00,000
China	60,000,000
Total	162,104,000

This total made up slightly over five percent of the world's males. Colleen couldn't believe the figures.

"So, all the spraying plus the Baghdad bombing removed only one male out of twenty from the world's breeders?"

"Or, for every male sterilized," Jonah predicted, "nineteen able-bodied males remain to throw themselves into the breach."

"Well, that's ridiculous," Colleen said. "Even with all the profit-taking, the cost was only $60 per person sterilized."

"Even sterilizing a dog in the US can cost more," Jonah said.

The UN World Health Organization agreed. They found no significant reduction in population growth due to the virus sprays. To reduce population growth would take much more serious effort. Colleen and Jonah came up with a goal for the new Population Control Institute:

Since the earth can only support one-third of the people alive today, the Population Control Institute will undertake programs intended to reduce the numbers of people on earth.

"Well, it's simple and we can count to see if we're winning," Jonah said.

PEOPLE PREDATORS

Colleen finished hanging her new motto on the office wall:

Humans must become their own predators.

She grew flowers and potted plants around the room and a redwood burl in a plate on her desk. It reminded her of the similarities of the DNA in the redwood tissue and her own. Her office window showed fog drifting through the tops of evergreen trees regrown after clear cutting.

Jonah returned from trying to ease the University's restrictions over having students sign up for credit work with the Population Institute. He stamped into Colleen's office, poured some coffee, threw himself into a chair, and glared at her. "Ass hole."

Colleen stood up and moved near his chair.

"Watch the coffee," he said as she slipped into his lap.

"Does this orifice have a name?" she asked. "Perhaps a civil servant in our new home town? Surely not me?"

"Dr. Harbinger, head of the biology department, Arcadia University's elder theoretician," Jonah said. "He called me a hammer-handed empiricist."

"Oh, the brute. But you are a hammer-handed empiricist, aren't you? I take it you talked to him about setting up the interdisciplinary teams for independent study projects?"

"Well of course I am, but he meant it as a criticism," Jonah said. "He gave me lecture number seventeen about the importance of students testing hypotheses derived from theory."

Colleen wiped at some spilled coffee and brushed hair out of his eyes. "Then you gave him lecture number forty three?"

"That's right," Jonah said. "I told him our theoretical knowledge is a hundred years ahead of our social practice and that the lag is about to kill us all." Jonah frowned. "It's hard to give somebody a lecture in his own office. Besides, full professors keep their hearing aids turned down."

"So what did you find out?" Colleen asked.

"We can set up any student projects we want for independent study credit, as long as the student's advisor will sign for them."

"Hey. You did good," Colleen said.

"Yeah, but now we have to meet some student advisors to get their support this afternoon. Are you ready?"

"You start 'em, I'll finish 'em?" Colleen asked.

"OK. You're the word weaver."

A dozen faculty advisors arrived at Colleen's headquarters late that afternoon, tired and hungry, not the best time to get approvals for a new program. She revived them with coffee, tea, and trays of cheeses and meats. A department chair said, "They must want something."

Jonah described the Population Control Institute as a hands-on, real-world effort to reduce the birth rate. "We have lots of technical information, but making things work in practice requires interdisciplinary help. We would especially like the assistance of students and faculty trained in the behavioral sciences: for the moment, psychology, sociology, and anthropology, ethnology, and maybe ethology?"

"Dr. Jonas, I'd rather think that we educate students," an English professor said. "One trains dogs."

"Yeah, and the military defines education as ineffective training," a bearded psychologist said. He delivered this blast without turning his head.

"Come on, you guys," said a large affable man. "Our new friends will think we're hidebound. Tell us about some of the projects."

"Good," Colleen said. "Some of you may have seen the research on changing attitudes in the viewing public through watching romantic situational programs?"

"Romantic situational programs? You mean soap operas?" the same English professor asked. Now she looked intrigued.

"Yes," Colleen said. "Soaps. Shifting of attitudes through program selection."

"Content analysis shows soap operas support the status quo regarding women's role in society: family; home; social and sexual relationships. The sponsors sell household products, beauty aids, feminine hygiene and drugs. The plots avoid controversial social problems and changing roles for women." Colleen stopped for breath.

"Since we're rapidly depleting our natural resources and can't feed our population. American women should lead the way in reducing the birth-rate."

"So you want to train scriptwriters?" the English professor said. "Educate them?"

"Yes," Colleen said. "We want to produce a series of programs that examine the negative aspects of having children. Today's soap opera programming ignores the effects of Downs syndrome, birth defects, mothers' weight gain, and other problems which consume parents' income and emotional energies. These problems take food and care away from the rest of the family."

The psychologist added, "Not to mention the emotional impact on family members. The birth of a child can cause depression, divorce, suicide. People should know their emotional and genetic risk profiles before they have children."

"Exactly. That's the kind of information we think women need," Colleen said. "We would like as many faculty and students from the health care field as possible, especially public health. We'll set up teams of students to write soap opera scripts so that the impact of having children is depicted more honestly. We also need to socially legitimatize the decision not to have children."

"I assume the teams will work on getting their scripts produced?" the English professor asked.

"Yes," Colleen said.

"Interesting," the English professor said. "Should I recommend students?"

"Please do," Colleen said. "Clearly, we need research on what's already been done, ethnic and social-class responses to existing programs. We welcome students from all disciplines if they have a contribution. And faculty." She looked pointedly at the English professor.

Jonah added, "There are other projects involving the mass media, so students interested in any aspect of news presentation through the media should inquire. We need sociologists and social psychologists for that effort."

"This brings us to the East Indian Government inquiry," Colleen told them. "We've been asked if we can develop a product to put in each Indian woman's hands the ability to sterilize men near her. Basically this involves the development of an easily delivered medium which will keep a virus molecule intact and viable until it encounters the mucous tissue of the male."

"We'll need biologists, chemists, cultural anthropologists and others for this project. The Indian Government has sought the help of the United Nations World Health Organization for distribution."

The faculty group asked questions, especially about how the plan might work in other cultures, for example, the US.

"For now, tell us if we can expect your support or participation in the various projects," Jonah said. "We need you to identify students, to recommend them, and sign their academic course cards giving them permission to take units of directed study. Each student will write a learning contract showing what she must accomplish to get credit. You can read the learning contract to see what she has agreed to deliver."

Most of the faculty advisors said they would help.

After the group left, Colleen asked Jonah, "Do you think they'll recommend students and sign for them?" She wanted to compare her perceptions with Jonah's.

"I expect so. Gives them prestige on campus."

40

EASY EXIT

Colleen found Jonah sitting in his office one day, staring at the wall, the telephone hanging from his hand. She hung up the phone. "Something bad?" she asked.

He nodded yes. "That was my mother. My father's cancer treatment is not going well. I'd better go see him."

Jonah disappeared. Two days later he returned to talk with Colleen. "I didn't anticipate how much my father would fear dying."

He stopped and thought. "No, he's not afraid of dying. It's the radiation treatment effects, the chemotherapy nausea, and the intravenous tubes and stents and catheters stuck in him, and worst of all, the pain. I listened when the doctor told him he couldn't save him from all the pain, especially toward the end. So my father sat there looking frightened." Jonah stopped in distress.

"There's more," Colleen said.

"He asked me to help him commit suicide if the pain gets too bad toward the end."

"Oh God," Colleen said. She put her arms around Jonah. "What are you going to do?" Colleen asked.

"I'm not sure yet." But he knew what he had to do.

The next day he visited representatives of the Retired Persons Association and arranged with their Board of Directors to commission a study of newer techniques by which older persons could end an intolerable life and avoid excessive pain. Jonah, through the Institute, would do the work and write the report to be delivered to the Retired Persons Association.

Colleen added a new objective to the Institute list:

Help return mortality rates to natural levels.

The project to help people end their life had not been anticipated, and Regis wanted all the legal bases covered. When he was satisfied he told Jonah to go ahead. Now Jonah put in motion the first Institute project that would recruit University staff and students. The time had come to select the first work team.

Jonah woke first on this important morning. He wanted to enlist aggressive, hard-working, inventive students for this project, and today he and Colleen would interview the candidates. He worried. Today tested the entire Institute model. If things didn't go well, they might jeopardize the whole Arcadia operation. He would hate to leave. He lived near water and boats and out of the city.

He chided himself for avoiding getting out of bed. He turned to Colleen and enjoyed her sleeping profile.

"You're getting prettier," he whispered in her ear.

"What?" she muttered sleepily.

"It's a sunny morning," Jonah said.

She turned her head toward him and opened the closest eye. "That's not what you said."

"I said, 'You're getting prettier,'"

She opened both eyes. "Really?"

"Yup," Jonah said. He brushed her hair to one side with his hand and stroked her cheek.

She felt between his legs. "You're not just trying to get me interested?" She sounded puzzled. "No, but I can take care of that." She began fondling him.

"This is a big day, you know," Jonah said. "If we don't get a good turn-out of student applicants, we won't have any chance of operating the Institute the way we've planned."

"That's true," Colleen said, still fondling him. Suddenly, she felt fearful. "I just remembered that I used to wake you this way every morning in North Carolina. Remember?"

"How could I forget," Jonah said. "You pampered me."

"We needed a daily sperm sample then. Oh my." She hesitated, then plunged ahead. "Do you miss the attention?"

"Yes," Jonah said. Now Colleen feared the worst.

"But I wouldn't trade my time with you now for those days," Jonah said. "You're more peaceful now. When you make love to me, you're more focused. You concentrate on enjoying yourself. It's reassuring and very satisfying."

Colleen breathed again. "Ooh. I'm glad."

Jonah started to get up. "Buy you breakfast?"

"Come back here," Colleen said. "The students can wait a bit."

MOTHER'S MILK

Because of his father's condition, Jonah decided to tackle the project of assisted suicides first. It surprised him that so many of the elderly he met were seeking a painless, safe, and inexpensive way to end their lives.

When Colleen and Jonah arrived at the University they went to the student advisement center. The faculty advisors had promised to assemble a dozen students with some biochemistry and pharmacology in their backgrounds. Jonah and Colleen met with them individually and then as a group, to lay out the goals and decide on the procedures for assisting voluntary suicide.

"How do you like them?" Jonah asked when the students left.

"Training, adequate. Enthusiasm, sensational," Colleen said.

During the next weeks, the students reviewed the pharmacology and anesthesiology literature. During a group meeting, they wrote their findings all over the blackboard.

To their surprise, sleeping pills and a cup of vodka remained the simplest approach. Intravenous potassium chloride caused loss of consciousness and death, but the patients couldn't self-administer. The cyanide pill of the old spy movies didn't produce the instant and painless death as portrayed in the films. Some of the newer psychedelic drugs with amnesiac effects produced a pleasant and quick black-out, but how could the patient obtain and administer these?

The students designed simple devices for the delivery of anesthetics, perhaps halothane or cyclopropane, into the lungs as a fast way to get the drug into the blood stream. Eventually, they assembled kits for several of these procedures, some to be offered for sale.

The elixir of choice, especially among drinkers, came to be called Mother's Milk. The recipe read:

From a pint bottle of vodka, empty one-third.

For flavor, add canned milk, sugar, cinnamon, nutmeg, vanilla to taste.

Shake well. Refrigerate until ready for use.

Open 40 of the strongest sleeping capsules available over a sheet of paper. Pour the powder into a small container until ready for use.

To use, place sleeping pill powder in mouth and wash down with as much of the vodka and milk as possible.

You will drift into unconsciousness. To be more certain, place a plastic bag over the head with a rubber band to exclude air. Hold open until you lose consciousness. You may wish to focus on pleasant thoughts such as your favorite place as a child until you lose consciousness.

The pharmacology team provided several alternatives in its report. In addition to Mother's Milk, the team designed several kits to deliver anesthetics to the lungs.

Regis arranged for production of three of these kits with equipment designed to deliver a high concentration of anesthetics into the lung's alveolar passages rapidly for fast black-outs. Those which worked best would be manufactured overseas, and sold through mail order, so Americans could buy them readily. The Retired Persons Association published all the project findings and provided advertisements and pamphlets to the public.

Twenty four hours after the project descriptions reached the Retired Persons Association, Jonah's phone rang. He sat at his desk admiring the team drawings of the anesthetic kits and the skill they showed. He felt pleased at the students' ingenuity.

"Mr. Jonas?"

On instinct, Jonah disliked the man's voice. "Actually it's Dr. Jonas."

"Yes, of course, Dr. Jonas. This is Dr. James Morgonstern, President of the American Medical Association. I've published in the field of anesthesiology. You may have heard of me."

Oh, oh. Big shot, Jonah thought. "I don't think so. What did you want?" Jonah felt angry and impolite.

"I've heard rumors about a project under your supervision that calls for prescription anesthetics. Have you a physician on your team?"

"No, we don't. I'm a PhD pharmacologist and qualified for research on these materials," Jonah said.

"Well, yes," the AMA president said. "But I'm sure neither you nor the Retired Person's Association wants the AMA to condemn your project?"

"What is your objection?" Jonah asked.

"Well, it was bad enough that Kevorkian can now perform assisted suicides. But your do-it-yourself kits give the public free access. That is unwise, and I think it is immoral," the AMA president said.

"How much money will AMA physicians lose if patients begin to cut short their prolonged illnesses?" Jonah asked.

"Are you trying to be insulting? I called you to attempt to negotiate."

"You have some suggestions?" Jonah asked.

"Well, at the very least, we want the drugs and kits to be prescribed by qualified physicians."

"We prefer to provide them to the general public, Dr. Morgan."

"Morgonstern. But anyone could take her life into her own hands, perhaps when she is psychiatrically depressed or otherwise incapacitated."

"I suppose that could happen," Jonah said. His hand tightened around the letter opener as he thought, You want complete control over everyone's life, don't you?

Dr. Morgonstern sighed. "I have already phoned the US Attorney General. You don't care if we also involve the Justice Department and the FBI to stop you?"

"I'd rather that you not do that," Jonah said.

The next day, an FBI agent arrived and asked questions about the Institute's operation and its first project. Colleen and Jonah called Regis who told them, "Don't worry. The FBI always tries to bully and frighten people." Sure enough, Jonah thought, I'm frightened and bullied.

NETWORKING

Twenty years earlier, while Regis and Colleen grew up, their father often took them into the wilderness. Colleen collected specimens while Regis fished. Colleen photographed while Regis hunted. No one expressed surprised when Colleen went into biology.

Regis, however, surprised his friends when he went into law. He showed a stodgy desire for justice and thought people should be fair. Of course, all these hopes would be knocked out of him by his training and early experiences as a lawyer.

Regis's partners realized that his success with clients came from his plain speaking, his serious concern with fairness, and his impatience with the sophistication of San Francisco, "The City."

He spent time, recently, as an organizer with the National Association of Riflemen. He felt they could contribute to saving the wilderness if he could focus the organization's efforts in support of the environment. NAR members felt isolated from society, their image ridiculed in the media. They saw businesses give and take away jobs, leaving the problems of white collar crime, urban drugs, and street crime unsolved. They felt unsafe, unprotected by the police, sometimes even preyed upon by corrupt police who would like to take away their weapons.

Regis wanted to reduce the antagonisms among the NAR, their media critics, and environmentalists.

At the moment, Regis drove to Arcadia State University. After he left San Francisco, he turned inland through the Clear Lake and Mendocino National Forest area. There he picked up two fellow members of the National Association of Riflemen to accompany him to Colleen's next meeting at the Population Institute.

"What's this all about?" Henry asked. "This better be good. You know it's pigeon season? Or have you forgotten about that sort of thing in your bigshot law practice?"

"You still bother with those poor little birds?" Regis asked. "They don't give you that much to eat. Why don't you guys ease up? They're getting scarce."

Henry's friend Gerald explained, "Well, he doesn't really go out for pigeons. He's still seeing that gal in Ukiah, so he needs every excuse to travel. By the way, didn't your sister win a science award the other day?"

"Yes she did," Regis said. "That's why I asked you to come with me. We've set up this new Population Institute, and we can't get a fair shake in the media. They won't mention anything about overpopulation in the newspapers or TV or magazines. All they'll talk about is growth and more growth."

"Sounds like us. The NAR has a bad name all over the country," Henry said.

"Exactly. That's why I wanted you to come along," Regis said. "You're going to be surprised to hear some of these people discover things we've been talking about for years."

"Yeah, but usually these people are pretty liberal," Gerald said. "Aren't our necks a little too red for them?"

"Maybe," Regis said. "They might get uptight if I introduced you as members of the National Association of Riflemen. But I'm going to introduce you as outdoorsmen, concerned over the loss of wildlife habitat."

"I'm going under cover." Henry laughed.

"Sounds funny, but I claim we all need each other," Regis said. "They sure as hell need us."

"Who is 'they?'" Gerald asked.

"A bunch of college professors who are tired of having their population worries ignored by the newspapers and TV. You'll hear from environmentalists, biologists, chemists, writers, computer types."

"You sure about this? I'm not convinced I have that much in common with those folks," Gerald said, trying to scuff dirt off his shoes.

"No, I'm not sure," Regis said, "but I want you guys to listen to what they're trying to do. Then if you have any interest in it, I want to try out the plan with the NAR. Try to ignore the do-gooders. Hear their plan to establish links all around the country and to get their own TV, radio, and newspaper coverage out to the general public."

"The NAR sure needs that," Henry said. "Hey, pull over quick."

Regis saw flashing lights behind him and turned off the road. A fire truck roared by, the people in back still strapping on clothing and equipment.

"Gimme, quick," Gerald said, pointing to Regis's cellular phone. He muttered at the mouthpiece in near English for some time.

"There's a fire on the coast," he announced. "They've called out all units, clear up to Ukiah."

"That's bad," Henry said. "Who did you talk to?"

"Rick's fire crew from the lake." Gerald rattled off a dozen other names and teams they both knew well.

Regis said, "Hard to hide anything from the local residents, huh?"

"Impossible," Gerald said, handing back the phone.

41

THE MEDIA

Colleen and Jonah carefully planned the meeting to which Regis and his friends came. Colleen wanted everyone who attended to share her anger at the media.

The group met in the new Institute office building rather than at the University, with the hope that someone might get out of hand. Colleen wanted the meeting to be hard hitting. She intended to recruit the faculty members for the media project.

The first batch of students who worked on Jonah's pharmacology project told him who to invite. Jonah asked the students, "Which faculty members have strongly developed social consciences, and are admired by their students?"

Jonah and Colleen first invited a white-haired, feisty professor of philosophy named Theo. They chose a dozen other faculty decision leaders from different fields. Jonah sought out the hot-bloods, articulate and anti-establishment. In the midst of these folks sat Regis and his red-necked hunters.

Colleen programmed Jonah to start. He still acted as front man with his "Honest John" speech and appearance. She would step in and dazzle them later, if needed.

"When this meeting is over, we hope to have a plan to create our own news outlet. I'm outraged at the lack of scientific information in the news. The so-called 'experts' who appear are usually businessmen speaking in their own interest." Jonah paused.

"You may have heard of the 'Growth Machine.' It consists of

commercial property owners, bankers, lawyers, accountants, realtors, chambers of commerce, local politicians, and the police. They attract outside investment to increase the area's population and drive property values up. Newspapers, TV, and radio stations benefit from more advertisements and subscriptions." Aroused, the sociologist had to speak.

"And at the national level corporations lobby and purchase politicians. They own the national news media and report everyone's wrong-doing but their own. They select and sanitize the news and blame the victims. You don't hear much about corporate crime. Then when they are through, they tell us, 'And that's how it is.' Actually, 'That's how it isn't.'"

"Atta boy." The group laughed and cheered.

"One franchiser of 150 newspapers requires the editorial policy to be, 'pro-business, pro-growth, and pro-development.'" Colleen said. "So, who will tell people the down side of growth, the bad effects?"

Colleen introduced Theo to talk about advertising.

"I wish to condemn the damage done to our future intellectual resources, the children. Television favors imprecise language, legitimates astrology, and new wave mysticism, and portrays scientists as impractical bores. But there's worse," Theo said.

"Advertisers violate every rule of logic by which we differentiate truth from fiction. How can children learn logical inference on a diet of TV half-truths and plain lies? Advertising destroys rigorous thinking and should be made illegal." An excited English teacher couldn't restrain herself any longer.

"He's right. Most of my students write in garbled fragments, and describe the world using fictional ideas taken from TV. My students cannot differentiate between propaganda and ordinary exposition."

The sociologist erupted again. "News corporations tell of crime in the streets, ignoring the fact that corporate crime does many times more damage to society. We are told that crime festers in ghettos because the inhabitants didn't know their fathers, or didn't go to school. They always fault poor people."

Henry whispered to Regis. "These people are pissed-off worse than we are!"

"Toldja," Regis nodded.

"A dozen books show how media news is invented by businessmen, reporters, and political newsmakers." Colleen decided they were ready. "So, I made an appointment with Burner Broadcast Company." The listeners snapped to attention. Jonah took over.

"Fred Burner gave copies of the first World Watch Institute environmental warning books to 500 of our national legislators. Yesterday, Burner told Colleen he admired her accomplishments and would put aside time to talk. We plan to ask for prime time to present our version of the news."

"That would be a breakthrough," said the English teacher. "How would we gather the news items?"

"The Internet," a student exclaimed. "The world wide web and e-mail," another said.

They soon had a plan for academicians, scientists, and students to assemble news stories. They hoped to break the corporate monopoly on "news." The news would expose corporate lobbying, white collar crime, and campaign contributions. The growth machine, advertising, marketing became vulnerable. Could they defend the environment? How exciting!

•

On the way back from Arcadia to San Francisco Regis and his hunters talked about the conference.

"So why did I come, and what am I supposed to do?" Henry asked. "By the way, I like those people."

"Me too," Gerald said. "They're feisty."

"Yes," Regis said. "Those people will oppose population increases. They want forests and wild life habitats in one piece; they're against overfishing, overhunting, and poaching; and some own guns."

"OK, we should support them?" Gerald asked.

"That's right." Regis mentioned the attacks on his sister. "Colleen and Jonah are honest and tough, and they'll make some people mad. I'd feel better if our people could keep an eye out for them. I know you monitor what's going on with radios."

"Well sure. When you're out in the country, you have to keep in touch," Henry said. "You think there could be trouble, and they might need help?"

"Yes. That's what's worrying me," Regis said.

"Well that's easy. Hold the next NAR shooting competitions up in Arcadia," Henry said. "We'll give tours of the Institute. Have Colleen and Jonah enter the NAR shooting competitions. Get Colleen to give a fire-eating speech on standing up for your rights."

"You guys are great," Regis said. "You think we could invite some of the tree huggers? They are doing good things in Anderson Valley. We ought to be on the same side as those people."

Gerald looked uncertain. "Maybe invite a few we know personally at first."

When Regis got to San Francisco, he called Jonah and told him about the favorable reception from his hunter friends. "They're members of NAR," Regis emphasized. "They recommended that you and Colleen join the NAR, and the local gun club. Go on their picnics and shooting meets. Get to know them. Tell Colleen to win another rifle competition."

"She wonders, why all the urgency?" Jonah said.

"Up there the back-country informal system is strong. People in their kitchens, vans, pickups, and campers, all own telephones, scanners, and radios. They listen to ambulance drivers, fire fighters, sheriffs and the State Police. They report the weather and fire hazards. It's a citizen's network, and you may not like them all. But, as President Johnson used to say, I'd rather have them inside our tent pissing out than outside pissing in."

"I'll try to explain that to Colleen," Jonah said.

"Hey," Regis said. "I really enjoyed seeing you two without my tie on."

Colleen and Jonah got big on the gun club circuit. Jonah had always wanted handgun practice. Some of the survivalists frightened them a bit, but being part of the communication network seemed worth it.

UNWANTED PREGNANCIES

"I'm still unnerved by what this contract calls for," Colleen told Jonah. "It says that the Indian Government, in cooperation with the UN World Health Organization, will pay us to provide them with a permanent supply of the Jameson virus to be made readily available to all Indian women. The words, 'all Indian women' startled me."

Jonah pulled a book off the shelf, thumbing to a dog-eared section. "Well, here's the background: From 1966 to 1977, Indira Gandhi served as India's Prime Minister and is known for her compulsory sterilization programs. The resistance she met left governments searching for other means to slow the population growth."

"They've made progress." Jonah said. "Where did the letter come from?"

"The Indian Consulate in San Francisco forwarded it."

"Good. Isn't that what we are all about?"

"Damn right," Colleen said. She ran the letter by Regis and the Institute directors for approval.

The India Contract seemed very straightforward. To obtain the students who would do the work, Colleen recruited only women.

"Why only women?" Jonah bristled.

"Males have a sterilization hazard," Colleen said.

"We've never cited that rule before." Jonah accused Colleen and the other women of sexism. They didn't seem to care. It was their turf, by God.

The women sequestered themselves in a lovely mountain retreat. They talked about the terms of the contract and their feelings about participating.

A very serious anthropology student announced, "I'm not sure about the ethics of another unwanted Western imperialist intrusion into Indian affairs."

A firebrand public health student responded.

"Not to worry. It's quite ordinary for the Indian population to talk about sterilization. Thirty to forty percent of couples are protected from unwanted pregnancy by vasectomy or tubectomy."

"That's right," a nurse said. "The Indian women came in greater numbers when they received a small incentive of money or a new sari after the operation. Sterilization is better accepted than we realize. It's our hang-up, not theirs."

Next, Colleen proposed her insight that humans must be their own predators, since humans could now elude most natural predators and overpopulate.

"I don't like the idea of being our own predators. I prefer a more passive role," said a dark-eyed, thoughtful psychologist. "Besides, it sounds like cannibalism. Why can't we try to relate to people's basic needs?" Later, she quit the group.

"The Indian culture also legitimatizes infanticide, and they usually kill the female babies," complained a feminist volunteer.

"How do we convince a woman to give up a function our culture defines as her most important contribution?" a sociologist asked.

Colleen wanted them strongly committed to the contract goals and continued to press them:

"Would you rather control who fathers your child or leave it up to the desires of whatever man you are with at the moment?" she asked.

"I'm not turning over that decision to any man," a Planned Parenthood advocate said. "Anyway, that's what goes on now, and it surely doesn't limit births.

"Would you prefer that men who run churches decide who have children and how many?"

They were emphatic. "No!"

"Aren't we willing to be guided by our various ethnic beliefs and social customs?" Colleen asked.

"No." they insisted. But after more discussion they decided those customs and beliefs currently do control their own reproductive behavior.

"Why not leave it to chance?" another asked. They didn't like that. Weren't university students supposed to use their minds to make choices?

Colleen tried once more. "How about genetic profiling to identify and reduce detrimental genes in the population? Then license the parents." That gave them trouble. They thought they should say yes, but didn't like the image. Fortunately for them it was impossible to administer.

After days of soul-searching, they came up with the guiding principle for the project:

Any woman in the world should be able to prevent
any man from getting her pregnant.

The virus should be distributed within a small piece of stable material, perhaps a crumb of bread, which could be crushed with the fingers to release the virus. The fragments could be placed in the male's nose, on mucous membranes of the genitals, or elsewhere. Placement should be simple, so that the man need not know it had been done. The virus-bearing crumbs should be readily available, preferably from other women, inexpensive or free if possible, not to be sold commercially.

Their soul-searching discussions and emotional struggles served to commit them to the project. For a few it became a sacred offering to their sisters across the world.

They had bonded, as Jonah noted, and men didn't belong. He wondered whether there might be some serious consequences to this later.

The Indian Contract team assembled six small and inexpensive virus propagation labs to assure an ever present starter quantity of the virus. Colleen also gave Jonah a dozen samples to squirrel away as they did when they left the Berkeley lab, just in case their situation got unstable. That's as close as they would try to describe the possibilities of a violent future: "unstable."

The whole team tested different matrices to carry the individual virus samples.

"That looks like dry spaghetti," Colleen said.

"That's what it is," the student said. "Then we check its shelf life under different atmospheric humidities and temperatures."

They tried specimens of dried cereal and powdered wood. They even tried bits of brittle styrofoam plastic. Then they tested them for viability over varying lengths of time.

Because the Indian Government wanted to start immediately, Colleen's team prepared the early report and collected samples of what they thought would be the most promising matrix materials impregnated with virus. Colleen's entire team of eleven women drove to San Francisco to present the preliminary report along with a small quantity of the granules to the Indian Consul.

They asked for and received a stirring rite of passage for this event, this outreach of a gift, perhaps freedom from bondage, to the world's women, their sisters everywhere. A World Health Organization women's representative arrived and gave a speech, as did the highest ranking woman in the consular office. They ignored the Consul, a male, who wasn't sure he approved of the contract, anyway.

When the women returned to the Population Institute, they prepared a simple description of the virus propagation lab unit and its construction requirements. They asked their computer friends to forward it to appropriate Web sites on the Internet and other repositories where it would be sought. The women wanted their contribution to live forever.

Next, Colleen found a request (almost a demand) waiting in her office from the National Women's Organization (NWO). The NWO came upon the plans for the virus propagation lab along with the description of the virus starter materials and the statements of feminist philosophy. The NWO people wanted to be involved. They immediately started building propagation labs in several feminist strongholds. The NWO asked for virus starter batches.

Colleen asked for approval for this disbursement of the virus at the next meeting of the Board of Directors attended by Colleen, Jonah, Regis, a San Francisco urologist, a gynecologist, and a public health PhD.

Colleen asked, "Do you see any reason we shouldn't give these people start-up batches of virus? Bear in mind, we've already provided the United Nations and the Indian Government with starter materials."

"My idea of a defense," Regis began, "if one were needed, is that the virus could be obtained freely in nature near the most recent spray operations. Viable amounts of the virus remain for some time in the soil and water of the region. We can't prevent them from getting those starter samples."

"True," Jonah said. "Now what if we are asked for starter samples by members of the American Nazi Resurgence Party?"

"They are males and fascists, two reasons not to let them have any," Colleen said.

"That's sexist," Jonah said.

Colleen glared at him.

"Let's argue that issue when it arises," Regis said.

They decided to provide the National Women's Organization with virus starters on the grounds that their needs paralleled those of the Indian women. Regis sent a contract for approval by the UN World Health Organization for disbursement to the NWO.

42

The [person] that rules the press, the radio, the screen, and the far-spread magazine, rules the country. Judge Learned Hand, 1942.

"Hell, you're just a couple of kids," Fred Burner said, as he rose from his desk to greet Colleen and Jonah. He looked them over openly as he led them to a work table alongside his desk. A tray stuffed with fruit, pastries, breads, and cheese, sat next to a perking coffee pot. Jonah was attracted to the clutter of reference books, two computer terminals, a rack of drafting pencils, straight edges, boat curves, and a sheaf of cross-section paper.

A blow-up of a racing sailboat covered one wall of the office. Lean and aggressive, Burner looked sixty, his gray hair losing its last bit of color.

"They tell me you guys had a rough time," Burner said. How did you get started with this Birnham Prize business?"

Part of Burner's charm lay in his accessibility. He didn't try to finish each sentence or idea. His voice tagged along after his mind which jumped curiously among questions about people, ideas, recent events, observations about himself, and assertions about the world. Colleen gave herself over to his agenda.

"You guys both went to college out of high school? I often wondered if that would have changed things for me. So what happened to get you so famous so fast?"

"I'm not," Jonah said. "The excitement started when I met Colleen at Berkeley."

"That's a good school," Burner said. "Ever meet any communists there?"

Surprised, Jonah tried to remember. "Maybe one. I'm not really sure."

Burner shrugged. "So you two met in school?"

"Yes," Jonah said, warily. "When Colleen applied for the Birnham Prize, some foreign groups wanted virus samples. We ran from them until we got picked up by Biosolz Pharmaceuticals. We set up the virus production system for them."

Burner turned abruptly to Colleen. "How do you like him talking for you? Is he a bully?"

Colleen hooted with laughter, charmed by Burner's intrusive curiosity. "He's terribly bashful. I asked him to speak for us for a while. I used to talk a lot until I found out people often lied to me."

"So what did you want of me?"

"Your admiration for World Watch Institute is well known," Jonah said, "as is your concern for the environment."

Burner nodded.

"We feel the media is failing to alert the public," Jonah said. "The growing population is rapidly using up fuels, soil, ground water, and lots more."

"No question about that," Burner said. "But some of your folks would shut us down. No more industry, trade, growth, profit. How do you feel about sustainable growth?"

"We're already overshot," Colleen said. "We have 6 billion people. We can only feed two billion over the long haul. Fifty thousand children die every day."

"Two billion?" Burner asked. "That's the carrying capacity? Somebody calculated that recently?"

"Resource and Energy team at Berkeley, and it still hasn't reached the general public," Colleen said. "That's our point. We need a news program."

"On Burner Broadcasting System?" Burner asked, laughing incredulously. "Who would sponsor you?"

"No one at first," Jonah said.

Burner stopped laughing. Finally, he asked, "What would you focus on?"

"Everything that is unfairly excluded from current broadcast news," Colleen said. "The need to stop the world's economic growth, to save natural resources, and to reduce the burning of fuels. Then there is the matter of gross misrepresentation of the US plight by our marketers, advertisers, reporters, lawyers, and accountants. They lie in favor of corporations."

"You don't blame politicians for any of this?" Burner asked.

"I don't," Jonah said. "They just vote the interests of their corporate campaign contributors."

"Probably true," Burner muttered. "It's a hard argument to fight." He pressed the intercom. "George, can you get my wife on the phone for me? " He asked Jonah and Colleen, "Could you come out to the farm? I want my wife to meet you. She's going to have to hold my hand if we do this. I'll be damned," he said, with surprise. "I haven't felt a flicker of uncertainty like that for a while. You're right, you know, about the media's control. No question about that. By God, I'd almost like to see their faces."

The phone rang. Burner poked a button and spoke. "Hi, love. I want you to meet some people, a couple. I like them. They sound just like you. It's important and you have to save us. I'm wavering. Got some time?" He waited.

"That's great," Burner said. "Why not call Seymour and ask him over for drinks?"

The half-hour trip to Burner's farm brought them to a rambling and unassuming house. He led them onto a screened-in veranda, a huge entertainment area overlooking the pastures, with the city in the distance.

As they sat down to talk, a tall balding seventy-five year old man dressed in a white suit joined them.

"Hey, thanks for coming, Sy." While Burner introduced Jonah and Colleen and described the Birnham Prize, Seymour carefully sat down, arranged his cane and his legs, and pulled out a cigarette holder a bit shorter than President Roosevelt's. He inserted an extra-long filtered cigarette and accepted a tall mint julep. Finally settled, he turned to his hosts with a smile and a twinkle in his eyes. He carefully inspected Jonah and Colleen.

"What a pleasure, Fred, to meet a celebrity scientist. I would come on a moment's notice." He addressed Colleen.

"Dr. Jameson, I read about your identification of the virus. Most impressive. You two have made a real contribution to population control efforts." After a bit, Burner intervened.

"I'd better tell them that you directed the federal experimental educational laboratories some years ago. What did you hope to accomplish, Sy?"

"We had the ridiculous idea that we could take principles and theories from the social sciences and apply them to improve education. But we were defeated from the outset. The teachers fought us tooth and nail, subtly, but effectively."

"Why did they fight you?" Burner asked.

"I think the instructors like their high status, dominant position in the classroom. They are admired by a few of the attractive young people. They can select variable amounts of pain to inflict with their tests. And lecturing is fun, or reinforcing, as Skinner would say."

"So teachers are petty and uncaring?" Burner waved at Jonah and Colleen. "These folks want to bring more academic research and information to the public through TV. Why should we, if academic institutions are self-serving like the rest of society?" Burner enjoyed a battle.

"Well, let's see if we can compare academics and businessmen," Seymour said. "Push that computer over here." Burner leaped to comply. He liked to say that he had a computer in every room of his house. Seymour rapidly punched his way into a library catalog.

"I'll call up some research studies on higher education. Here are two I know about: 'How College Affects Students,' and, 'Research Fraud in the Behavioral Sciences.' The first book summarizes research on how colleges affect student learning. It shows few differences from one school to another. Claims of superiority are not supported. Now, have you ever seen studies by an American auto corporation saying that their products aren't measurably different than other auto manufacturers?"

Burner laughed. "Yeah, but that's different. Those guys are trying to make money, sell more of their cars."

"Exactly so. That's why academic research is more reliable than information offered by capitalistic corporations." Seymour leaned forward and said, "American academics aren't perfect."

Jonah and Colleen sat spellbound. Like fish in an aquarium unable to see the water in which they swam, Jonah and Colleen had seldom heard the dynamics of the academic environment discussed. Seymour continued in a high dudgeon.

"Academics don't seem to understand students and aren't comfortable with them. Popular teachers are resented by their peers. They are a threat and may not get tenure or promotion."

"Are you serious?" Burner asked.

"Yes. College faculty know even less of the science of teaching, but they insist they are experts," continued Seymour. "We've proved beyond any doubt that a student with a text and a set of quizzes can learn more on his own than listening to a lecturer. A psychologist named Keller showed how to do it, but the faculty wouldn't use it, even though the students learned more."

"On the other hand, at the big universities, the faculty members work hard. The media attack them, and corporations hate them because the academics try to tell the truth. They occasionally get caught up in fraud, but ordinarily, they are trying to find the truth. Corporations, and media advertisers, by contrast, are trying to make money, and they lie, cheat, and steal to do it."

Fred Burner held up his hands. "OK, OK. Jane," he called. "You gotta help. Come in here."

Burner's wife led the discussion back to the news project. Shortly, she was as enthusiastic as Colleen. Burner watched as his wife and Colleen made plans. Tell the truth during newscasts? Unheard of. He laughed helplessly at the thought of it. "Look at them. They're going to get us in a lot of trouble," Burner told Jonah.

"He really agreed to give you time?" Regis couldn't believe Burner would do that.

"He'll start us at five to five-thirty p.m., every day," Jonah said.

"That's a lot of money," Regis said. "This is out of character for American corporate executives."

"I think he remembers when he was a hungry house painter," Jonah said, "and he enjoys his wife. She and Colleen are talking like old friends."

"What will you do for news gatherers, script writers, video footage, all that stuff?" Regis asked.

"Colleen plans competitions for teams of college students to write 10, 20, and 35 minute stories on any topic not acceptable to ordinary corporate newscasts," Jonah explained. "Most often industrial misbehavior that isn't adequately reported. Like Rocky Flats burning a metric ton of plutonium and not telling the Nuclear Regulatory Commission. Someone should be in jail for that. Also, for the time bomb of leaking nuclear wastes at Hanford."

"The students will enjoy that," Regis said. "I'll have to get up to Arcadia more often. Sounds like fun."

SOCIAL UNREST

Colleen and Jonah were too busy developing the Institute to notice that an election took place in November. The Republicans won a majority in both houses. The Business Lobby urged large subsidies for production of coal and nuclear energy. "We need stable taxes, reduced welfare programs, reduced health care programs, and reduced social security payments."

Older Americans and other poor people began demonstrating and became unruly. Historians warned that insurrection, even rebellion, may follow disruption of people's housing, income, food, jobs, and family ties. The new year started with much confrontation in America.

The abortion issue heated up when a priest counseled a girl in an abortion clinic to withdraw and have the baby. Destitute after she gave birth, she sued the priest and the church for child support. Of course, the priest said he was not the father, but the court awarded monthly payments to be made by the church of the priest's parish. New mothers who had been counseled away from abortions now clogged the courts, asking for money from the church.

43

QUALITY CONTROL

Burner Broadcast System: Jameson Television Newscast #1
A well-known consumer advocate introduced Colleen to the audience.

"Tonight's broadcast introduces a new series prepared under the direction of a recently honored scientist, Dr. Colleen Jameson. She won the Birnham Foundation Prize for her work with viruses."

"Dr. Jameson, you work as a biological scientist at a university. As I understand it, you are paid to do research and to write articles about your findings."

"That's correct."

"Would you tell us about your award-winning discovery, and how you informed others of the results."

"The discovery was more of a chance encounter," Colleen said. "A sperm sample showed excessive cell destruction and I found a new virus in the sample.

"You wrote about your work? Was it exciting?"

"Wildly so for me. I wrote up the results and sent them to the *Journal of Viral Taxonomy.*"

"Let's ask the editor of the journal how he made the decision to print your article. Dr. Wyndham, what precautions did you take before publishing Dr. Jameson's article in your journal?"

Dr. Wyndham appeared on the screen. "I sent it out for review by other virologists. We cut author's names from the article. That way, reviewers can't be influenced by a writer's reputation."

"Why is such care important?" the interviewer asked.

"Your articles show other virologists the accuracy and importance of your work. You can't keep a research job without publishing."

"Who approves the articles?" the interviewer asked.

"Researchers most familiar with the subject."

Colleen intruded. "Dr. Wyndham, surely someone raised questions about my work?"

"Oh yes. Some had doubts, but if you were right we wanted to report your findings. Then others could test and verify them."

"What happens when information is incorrect, or even fraudulent?" the interviewer asked.

"Sometimes all hell breaks loose when readers think something is wrong," Wyndham said. "Researchers have lost grants and jobs when they made mistakes, or misrepresented the facts, or took ideas from others. Some plagiarists had their degrees revoked."

The interviewer turned back to Colleen.

"Dr. Jameson, tell us about your concerns with current television newscasts. What led you to want to do this new series?"

"Television broadcasters can present information without any scientific check for accuracy. You listeners know this when you refer to advertising as 'hype,' 'hard sell,' or put more plainly, telling lies to sell products."

"When advertisers increase company sales, they are rewarded financially for lying, while research professors may be fired."

On hearing this last statement, an advertising executive having a drink with his legal adviser exploded. "How do we stop this crap? This is an outrage."

"Aren't you a bit touchy?" the lawyer laughed. "If you can prove she's lying, we can sue her. Want me to work on it?"

The announcer asked Dr. Wyndham if he had any other comments.

"Yes. For a century, university researchers provided American businessmen with information leading to huge profits. Those profits went to corporate investors. The researchers should take those profits to finance their own research facilities and programs."

"Thank you Dr. Wyndham. Our viewers should look for Dr. Jameson's next broadcast this time tomorrow. The topic is corporate crime. You should find it interesting."

The program ended.

PATRICK'S PATRIOTS

An event with wide news consequences occurred while a cable-TV public access team routinely videotaped a Florida county council meeting. The council members sat on elevated chairs around a podium. This forced supplicants to look beseechingly upward.

As the camera panned around the room, it showed windows on one side, and two doors opposite, where a security guard sat. A dozen persons witnessed what followed.

An earnest local lawyer and the president of a development corporation completed arguments to the council as to why certain land should be condemned and rezoned for a golf course. Rumor had it that when the developer completed the golf course, he planned to build a hotel next door.

Barely noticed, an old man in work-stained overalls limped into view and approached the lawyer and developer as they testified to the Council. When they saw the old man, they blanched and tried to retreat. The old man shouted:

"First, you tried to buy my land from me for next to nothing. Now you're going to get it condemned and throw me off. You've paid these council crooks to rezone it for you. Well you aren't going to get away with it." He pulled a 9mm pistol from his overalls, and shot both of them twice at close range. The lawyer and developer fell writhing to the floor. The old man turned to the audience.

"Everybody keep your seats. My name is Patrick Stevers, and I'm going to tell my story." A woman close to a door ran out. The rest remained frozen in their chairs.

Patrick moved among the council members. "Don't try to shoot me," shouted Patrick to the security guard. He waved his gun. "This is one of them seventeen shooters. Got plenty left for the council members here if there's any trouble." Patrick's clear voice spoke with conviction. "Get all of this," he said to the camera crew.

Patrick described how disaster overwhelmed him. His father gave him the farm, and his sons took it over a few years earlier. He described the efforts by the developer to buy them out, and the expensive legal struggle that ended with condemnation and forced sale of his farm.

"I asked around to see if I had a chance with this Council. But I found out that you five, he pointed with the pistol to a bloc of men at one end of the table, always vote in favor of every development that is brought before you. Five to two. Five to two." He waved to the audience. "Just try to find a different vote pattern. You bastard councilmen vote for each others' projects. You are criminals taking money to vote against the interests of the voters you represent." As he talked he started swallowing pills he took from his pocket.

"If we didn't have enough trouble, my wife got cancer last year, probably from the pesticides we overused, at the direction of the agriculture extension agent. So now they are cutting Medicare, and I don't have money for her bills."

"The loan is overdue to the Farmer's Bank for the fertilizers they said we needed. Now, I'm pretty sure we over fertilized."

He finished swallowing pills, reached in his overalls, and pulled out a bottle of whisky. He took a long drink.

"And now, the government is going to take away my social security payments?" He looked around at the audience. "You have to tell them this isn't fair. Where is that miserable excuse for a newspaper reporter?" Hands pointed to a now crouching figure.

"Will you report this truthfully tomorrow?" The reporter nodded frantically.

"Patrick, this is the sheriff. Put down the gun," said a uniformed man who stepped inside the room.

"Don't be silly, sheriff. Stay right where you are and keep your gun in its holster, or I'll shoot up the room. Then you can explain to the viewers of this videotape how you caused that." He took another long swig at the bottle.

"Since you are here, sheriff, I want the people to know that you don't investigate these rich bastards, these developers who steal our money. And the County prosecutor never takes them to court." Patrick's speech slurred and his head nodded. He took another long swig at the bottle.

"Funny thing," he mused aloud. "My wife used to be happy. Then she got older watching all those damned advertisements telling her she's gotta get the wrinkles out of her face and lose weight so she can wear the new fashions. What the hell are they doing to us? Why do we hafta keep up with the goddamned kids as we get older. Everybody's gotta have new cars, new furniture, new hair shade, new razors." He looked up, angry. "Why is my wife trying to dye her hair? Salesmen tell us everything old is worthless, including us.

Hell, the older the car that still runs, the better the car. Those advertising bastards make us ashamed to look in the mirror."

He shook himself and turned to the two Council members who voted against the five-man bloc. "You people are OK, and I wouldn't hurt you. Don't worry." He turned balefully to the other five, his speech more slurred. "But you bastards, won't see your children again." Patrick then fired a dozen shots at the five men. Some collapsed where they sat and some fled toward the windows to their side. He shot them all.

Patrick drunkenly blew smoke from the barrel, squinted at the sheriff, and shouted, "Didja keep count, Sheriff, how many shots are left?" Patrick took another drink of whiskey, ready to collapse by now. With one last effort he said, "Where is that piss ant reporter?" He squeezed off a last shot at the fleeing man and missed. "Shit," he said, and slid from his chair unconscious.

As Patrick hit the floor, pandemonium broke out. Ambulances took the seven people to emergency. By the time Patrick arrived at the hospital, the pills and alcohol suppressed his central nervous system enough to kill him.

News reporters learned that Patrick had read the Retired Persons Association publications which told him how to use pills and alcohol to end his life. The commercial TV news showed excerpts from the videotape, selected to portray Patrick as crazed.

However, Burner Broadcast System played the complete copy of the tape on five to six p.m. news broadcasts the entire week. The commentator went through the sequence of events and explained the issues which Patrick raised. Clearly a harassed and wronged man, Patrick defended himself without shooting any policemen. In most circles, he became a folk hero. Bands of followers formed, called "Patrick's Partisans." Distributed by rental shops, the videotape did a lively business.

44

THE BUSINESS OF AMERICA

The second BBS Newscast began with a prominent UC Santa

Barbara sociologist who introduced Colleen.

SOCIOLOGIST: "Tonight we begin a series of broadcasts presented under the direction of Dr. Colleen Jameson. She recently won the prestigious Birnham Foundation Prize for her work on viruses."

"Academicians have provided the research for this program showing business and industrial leaders at the height of their corruption, acting against the interests of their fellow humans. Dr. Jameson will explain." The cameras turned to Colleen.

COLLEEN: "Good evening. The overriding problem scientists failed to solve for the last half a century is the dangerous growth of the human population. The corporate community makes a great deal of money from population growth. However, they aren't likely to tell us that current business practices kill 50,000 children every day. They cause some portion of the six billion people now alive to die of starvation. Some will die sooner of industry induced illnesses."

"Let's look at well-known misdeeds of three industries: petroleum, automotive, and pharmaceutical. In the past they violated government regulations so often as to require government intervention to protect the public."

An exasperated viewer, the head of a petroleum corporation, chewed on his cigar and complained to his aide: "Can't we get her for libel or slander, or something."

"For what, sir? Mentioning the government intervention?"

"No, Goddam it. For saying that we violated government regulations more than other industries."

"I'm afraid not, sir," said the newly graduated assistant. "I read the study, from the 1980s. It showed the violations, chapter and verse. Bunch of sociologists did it."

"Goddam it. I don't care if it's true. We've got to stop her," the boss exploded.

Next, Colleen introduced a petroleum specialist who had predicted oil field depletion rates. She described careless petroleum operations while the screen showed pictures of pollution from petroleum based industries.

PETROLEUM SPECIALIST: "The oil industry includes the largest corporations in the world. Their incomes are greater than those of most countries where they operate. Their tanker ships have more tonnage than any navy and they own whole cities. The U.S. Secretary for Energy once reported that he couldn't force U.S. refineries to produce more gasoline or heating oil because that might cause the oil companies to withdraw oil supplies from the United States."

"These transnational corporations concealed from you that US oil is nearly gone, used up by autos. They tragically wasted this non-replenishable fuel."

Colleen then introduced a colleague of Ralph Nader, a safety engineer who talked about automobiles, with background videos showing fiery automobile crashes.

SAFETY ENGINEER: "One of the frequent misrepresentations in the auto industry involves safety. The so-called public service announcements regarding damage done by drunken drivers is little more than a smoke-screen to hide the failure of industry to improve automobile safety. More money has been spent on cosmetic changes than changes designed to reduce highway deaths and injuries." Colleen's face appeared on screen.

COLLEEN: "Automakers' allies destroyed the US mass transit facilities 60 years ago. Without consulting the public, they bought and dismantled U.S. rail transportation systems that allowed safer and less polluting movement of people and material."

"Congress did nothing. Federal prosecutors did nothing. The auto destroyed rational community design, caused pollution and health hazards, and ruined large portions of the country's arable land."

The screen then showed footage of pharmaceutical corporations, with scenes of pills and more pills.

COLLEEN: "Careless proliferation of unneeded drugs endangers the health of the public and is actually causing the selective breeding of strains of microbes unaffected by the drugs. The new strains are more virulent and dangerous than those which the drugs were originally designed to fight. If we do not stop these practices, the new and murderous strains may become the predators which solve the human overpopulation problem through plagues more deadly than any from the past." The Santa Barbara sociologist returned.

SOCIOLOGIST: "Small businessmen have done extensive damage. Local chambers of commerce continually press for growth. This causes over-crowding, congestion, loss of soil, loss of arable land and water for agriculture, excessive burning of fuels, and release of pollutants into the air, water, and land. And all of this is done just to provide a profit for a handful of developers who will then move out of the areas they've ruined." Scenes of eroding agriculture land and city congestion accompanied the message.

Colleen introduced a woman from the Retired Persons Association against background videos of elderly Americans protesting in the streets.

RETIRED PERSON: "The media's attention is now focused upon a youth culture," she said. "Older people are devalued, their achievements, their lifestyles, their very selves rejected. We especially invite the older viewers to join groups, to share problems, and to validate the sort of existence which faces you in your later years. Tell us your special problems so we may assure you get a fair hearing."

"Locate the Grey Caucus, the Silver Legislators, and the Retired Persons Association, and help prepare programs and position statements for our broadcasts and for legislators." Colleen returned to the camera to summarize.

COLLEEN: "Many of these broadcasts will try to turn legislators' attention away from corporate campaign contributors and back to the people most needing help."

"Lobbyists beware. You threaten representative government in America. A corporate dollar given to a publicly elected legislator is a dollar given to subvert democratic representative government."

"In closing, let me challenge all students, young and old, to join others and describe problems which have not been reported in the past. You may communicate with us through your school, or directly to the Jameson Population Institute." The phone number of the Institute appeared at the bottom of the screen.

"Our students may be America's last remaining reserve of honesty, social concern, idealism, and care for living things. We will broadcast studies that bring to the public's attention information needed for sensible social planning. We will announce prizes for the most socially useful programs."

Colleen provided an 800 telephone number, e-mail and web-site address, and ended the broadcast.

As she left the studio Jonah clapped her on the shoulder. "I really liked that." Colleen wondered how many corporate executives she antagonized.

45

Washington is corrupt to the core. A. Etzioni, Sociologist

CROTCHETY OLD FOLKS

In January, when the new Republican majority took over Congress, over 35 million retired persons in the US depended heavily on social security payments. When the proposed cuts first hit the news, America's elderly cried out in rage.

The national Retired Persons' Association fought back. They prepared reports for "The Front Burner," as some called Colleen's daily newscast on Burner Broadcasting. The RPA Chairperson kicked off the series.

"Good evening from America's senior citizens. We join you tonight in the spirit of Montaigne who said 'I speak truth . . . as much as I dare; and I dare more as I grow older.'"

"As you may be aware, the Retired Persons' Association asked a number of our legislators how to improve the negative image of senior citizens in the media. We asked how we could regain control over the content of our newspapers and magazines, and our radio and television programming."

"Here is what we learned from a prize winning TV writer who is a TV Writing Institute director." A jocular, angular man's face appeared, animatedly explaining to an older interviewer:

"Why can't the public exercise control over TV programming? Despite a popular misconception, TV isn't a communications medium, it's a sales medium. TV sells us, the audience, to the advertisers, just like a slave auction." The speaker waved his arms expansively.

"The bigger the audience, that is, the more of us sold, the higher the price. That's what TV ratings are about. Then, programs are shaped to carry the commercials which bring in advertising revenue and pay TV salaries. TV programmers are sales experts, not communication or entertainment experts."

The Retired Persons' Association interviewer returned to the screen.

"Why, you may ask, do we allow our programming to be selected and shaped by corporate salesmen? Why not buy our TV programming directly?"

"It turns out that we can buy it directly. The legislature can remove business deductions for corporate advertising. Those monies belong to the voters. Why let advertisers spend our tax dollars to select our TV content and push their products?"

"So we asked legislators to eliminate advertising deductions, but they said they couldn't."

"'Why not?' we asked."

"They said businessmen would quit paying campaign expenses, and told us, 'We can't get elected.' They all agreed."

"But, we asked, isn't this just buying of votes?"

"The legislators told us the law allows contributions from Political Action Committees. Also, corporations can pay legislators honoraria, or fees for speaking engagements. Since some legislators have made more money from business honoraria than from their salary, some call these fees 'dishonoraria.'"

"To correct this situation, the Retired Persons' Association has identified the top fifty corporate contributors of campaign funds. We ask that you help us convince the corporate directors on the list to stop all contributions to legislators."

"Please talk about boycotting these businesses to your friends. Carry this message to your churches."

"We thank Burner Broadcasting for time to bring you this message. We will keep up the pressure until Congress removes advertising deductions. This is our Freedom From Advertising bill. We'll push to make all business and industry campaign donations illegal. That is our Free Elections bill."

"Now, here are the 50 largest rogue industries. We demand that they stop corrupting legislators with money for any purpose."

The RPA Chairperson read the list of corporations twice, and gave the amounts each paid to legislators.

The broadcast closed.

POLITICAL STRATEGY

The complaints of the Retired Persons' Association and the 35 million people they represented created great political interest. The "Goddam liberal President," as the conservatives called him, wanted to harass the new Republican legislative majority. That majority included a number of the religious right wing who hoped to insert their economic and moral beliefs into the law of the land. But the fiscally conservative Republicans wouldn't share the power.

The President gathered several close advisers for a strategy conference. The Democratic Party chair attended and told them political changes were afoot:

"The Republicans stirred up a hornet's nest by trying to reduce social security payments. Retired peoples' demand to eliminate campaign contributions and advertising is truly a radical shift."

"Except that everybody knows the Political Action Committees contributions are a scandal and should be eliminated," the Commerce Secretary said.

The Democratic Chair shrugged. "The National Women's Organization joined hands with the Retired Person's Association to support the bills. They also made the veiled threat that the NWO will distribute the Jameson virus to most American women. I'm not sure what the NWO hopes would happen. To allow women to sterilize their sex partner? Now that's structural reform."

"Can we get a bill passed banning PACs?" the President asked.

"If Democrats thought it would pass, many would join the few Republicans. I think it's possible," the Commerce Secretary said.

"You agree?" the President asked the Chair.

"Yes, it might have a chance. The legislators hate sucking around for campaign funds."

"Which Republicans would oppose the PACs?" asked the President.

Most came from the religious right, which excited the President.

"Let's jump on that. Go for a coalition."

CAST THE MONEY CHANGERS FROM THE TEMPLE

Two days later, an appointment was made discretely with a prominent televangelist, Philleus ("Bat") Masterson, who once ran for President as a Republican. The President, the Attorney General and two cabinet members met with "Bat," who couldn't imagine why he had been asked to attend. The President spoke.

"Mr. Masterson, I'm glad you could meet with us on short notice. How should we address you? Is my information correct that people call you Bat?"

"Yessir, since my youth. I'm comfortable with it. I don't use the name with my television audience. It might seem a little informal."

"Well, Bat, would you oblige us with a prayer to start the meeting?"

Bat barely concealed his astonishment. Then he grasped a hand on each side of him, and threw down his head. "Yessir. Dear Gawd, let us meet together this day, under your guidance and with your blessing to help us solve whatsoever problems we undertake. Amen."

"Amen," the group muttered. Bat looked around the table and recognized only the President.

"Thank you, Bat. You are probably wondering why we invited you. How close are you to your fellow Republicans in the legislature? For example, Senator Thruston of the Banking Committee?"

"We've known each other for a while. I met him at seminary. I know him pretty well."

"Yes," the President said. "Don't you also know Roberts of Ways and Means? And isn't Westmore a distant relative?"

"Well, that's right. Second cousin, actually."

"I'm interested in your opinion as to how well you think the Republican majority will do in congress during the next few months," the President said.

"Well sir, I reckon they've gotten themselves up a stump. They want to reduce the national debt. But everybody knows that Republicans and Democrats both have increased the debt about the same, no matter who occupied the White House." Masterson looked serious and continued.

"I don't think my fellow Republicans like the anti-PAC bill either."

"How do you feel about it?" the President asked.

Bat hesitated, wondering how candid he should be. "Frankly, Mr. President, I think it's needed. You might assume that since I raise money on TV, I would approve of PACs. But I ask for money for the ministry. These PAC groups buy votes for projects which couldn't get through congress otherwise. I don't think they are in the public interest. I don't expect a legislator in my district to represent me if he accepted a hundred thousand dollars from some businessmen. No sir, I don't think it's right." Bat finished in some heat.

"Can you think of anything we could do?" asked the President. "I'd like to offer bipartisan help to Congress, if I knew what would be good for the people. It would be good for my image next election time if people saw me try to help the legislature, and not simply act as an obstructionist. Frankly, that's why I called you here."

"Well, Mr. President, that sounds very sensible. As you may

know, when I ran for office, I advocated several measures to deal with the national debt," Masterson said.

"Tell us what you would do," the President said.

"Well, sir, I'd forgive it." Masterson took a deep breath and said, "It's the Christian thing to do."

The President leaned forward. "Just how would you go about forgiving the national debt?"

"Well, sir, I'd start by defaulting on the largest bonds. Anybody who ever bought a million dollars worth of bonds deserves to lose it."

The listeners sat stunned.

"How would you approach that, Bat?" asked the Secretary of the Treasury.

"Well, sir, I've given it some thought." Masterson said he would default on different bond issues and Treasury bills; he would lean on the Fed; and, oh yes, he would issue new currency. If the armed forces could regularly replace the military currency in occupied countries overseas, couldn't the Treasury Department do it in the U.S.? It would stop the money launderers, drug lords, counterfeiters, and other such.

Masterson waxed eloquent. Forgiveness of the national debt became the cornerstone. Wall Street was just a bunch of croupiers serving investment gamblers. He'd shut it down. When Masterson ran down, the President had one more question.

"Bat, would you consider public service? Giving your time to the government? Accepting a job with us?"

"Well of course I would, sir," he said, snapping to attention.

"The Treasury Secretary, here, needs to take a back seat for a while. He would be available, of course, for advice as you need it. But perhaps it is too much to ask."

"What did you have in mind, sir?" Masterson asked, beside himself with intrigue.

"Well, Bat, I wondered if you'd consider serving as my acting Secretary of the Treasury?"

"Sir, you're serious? You're not joking with me?"

"Not at all, Bat. I have wondered whether a Christian approach could improve the running of the government. We've tried almost everything else." He looked piercingly at Masterson.

"Well, of course. I would be honored to help, Mr. President." Tears welled up in Masterson's eyes. He had never really expected to apply his beliefs. Stranger things had happened. Remember how Reagan turned out to be a fundamentalist, and no one guessed? "Yessir, I'll help. When would you want me to start?"

They talked late into the night. After Masterson left, the President looked around the table.

"What do you think?"

The Treasury Secretary said, "Pass me that bottle of Scotch."

The Commerce Secretary chortled, "I think it is inspired. At best, he might arrange for the loss of the banking fortunes of several dozen of America's finest conservative dynasties. At worst, we can fire him."

"I want to be there when he announces plans to change the currency," the Treasury Secretary said. "America's money magnates are in for an exciting time." He turned to the President with an appraising look. "Sir, I didn't know of this cruel streak in you."

"Oh, yes. I'm afraid so," the President said. He had a new thought.

"Say, how would you feel about finding a Christian Scientist for Secretary of Health? Wouldn't that keep the American Medical Association out of our hair for the next two years?"

"Sir, stop this madness," said the Commerce Secretary, still laughing.

"Well, think about it. The Republicans may have shoved aside their radical right Christian minority so far. Why shouldn't we give the Christians another chance? Bat and a Christian Scientist Secretary of Health would offer programs that would incapacitate the Republicans. Why are you all laughing?" the President asked. He went on.

"Let's make sure these Republican financial lions are forced to brawl publicly with their Republican Christian brothers. Book 'em for the Washington, DC Colosseum for the next two years. We can fire Bat at some point, and rescue everybody just in time for the next elections."

"I'm a little nervous about giving these simpletons cabinet posts," said the Treasury Secretary.

"No problem," said the President. "This is just a generous offer by the Democratic administration to provide bipartisan assistance to the Republicans in the legislature. Make the Republicans turn down my appointments of their own party members. I want the American people to see what both sides of the Republican party are capable of."

The President asked the Commerce Secretary, "How about enlisting that isolationist senator June Borah? We could let her run the Commerce Department. She'd shut down the US borders. That would take care of NAFTA. Right?"

The Commerce Secretary couldn't stop laughing.

46

THE NATIONAL WOMEN'S ORGANIZATION

Colleen dragged Jonah home to watch the National Women's Organization broadcast on the BBS. An older, angry NWO Vice-President began the talk.

"The National Women's Organization presents tonight's broadcast. We wish to offer our support to older Americans whose social security benefits are threatened. Women, especially, need the benefits."

"Our society thrusts women into care-giving roles. Women raise children, and also are expected to care for older family members and relatives. Over half of these women, however, must work outside the home. Outside work and loss of social security benefits reduce the care that women can provide." The announcer turned to a nearby panelist.

"Tell us about the psychological impacts on the recipients of this care," she asked. Another worried looking woman spoke.

"As you can imagine, the elderly resent being forced upon their children, relatives, or informal volunteers for care. They prefer the support of public care, as an entitlement to services, rather than as a gift or charity."

"The elderly didn't choose to become dependent. As Americans grow older, they become poorer. They are pushed out of the labor market, with or without pensions. Any social service funds assigned to them become less through inflation. And, inflation is a built-in accompaniment of demand from business for continuous economic growth."

Suddenly, an anxious voice interrupted the speaker:

"I'm afraid so. There's a bomb threat, to go off in a few minutes. Every one must leave the building." The operator panned the camera away from the speakers panel.

Not at all finished, the NWO announcer shouted, "Get that camera back here. No practical joker is going to stop this broadcast. Turn that tripod around." After muted argument, the announcer's image returned to the video screen. "That's better," she said, squinting at the camera, her jaw set. "Where were we?" She located her place in the script, squared her shoulders, and continued.

"Our culture pushes women into another impoverishing cycle. First she marries, then has children, and too often she divorces. She finds herself older and much poorer through this process. We must do what we can to help women avoid it."

"For this reason, the National Women's Organization has started a project that will place in the hands of every woman the ability to prevent pregnancy. We will make the Jameson Virus available through the mail at a very low cost to any woman who wants it. We recommend that women resist the urge to have children. Our impulses lead too often to pregnancy, impoverishment, dependency, and political powerlessness."

"She's not willing to accept second class citizenship," Colleen told Jonah. The announcer continued.

"America's legislators must free themselves from the influence of the corporate Political Action Committees. Our senior citizens correctly demand an end to this blatant bribery."

"We send tonight's message to the U.S. Congress in the strongest possible terms. America's women support the demands of the senior citizens for continuing social security payments, and elimination of the campaign contribution to finance elections."

"We thank Fred Burner for making this broadcast possible free from the emotional pollutants of advertising distortions and business deceptions. We suggest that television's marketers and advertisers go find honest work. Good evening from the National Women's Organization."

Off camera, the NWO announcer snorted. "Men? Bombs? Bluster." Sure enough, it was a false alarm.

Whoever, directly or indirectly, corruptly gives, offers or promises anything of value to any public official . . . with intent . . . to influence any official act, is guilty of bribery or attempted bribery. Federal Criminal Code.

At the President's request, Bat Masterson's first act took him before a joint meeting of Congress. As a Republican member of the Democratic Cabinet, his staff prepared a bill that would make Political Action Committees illegal and would define contributions to a political candidate or office holder as bribery. The President's writers helped Masterson write the plea.

Without mentioning God directly, the speech frequently referred to driving the money changers from the temple and improving the morality of the House and Senate. To everyone's surprise, the speech got good reviews even by liberal columnists:
"Sensible, Long Overdue, Proper Application of Moral Suasion, Inspired."

Most Republicans argued against the bill. The Advertisers Association went on overtime trying to discredit Masterson's arguments. The national Chamber of Commerce and U.S. corporations fought vigorously against it.

By contrast, most church leaders supported Masterson. He went on national TV for the President and made gentle fun of the corporate caterwauls:

"Seems businessmen don't trust the citizens to make their own political choices. Businessmen demand the right to spend millions of dollars to influence how our politicians vote." The more the media protested, the better the bill looked to the working public.

EVAN

A fog blew across Humboldt Bay and drifted through the Arcadia University campus, a typical winter's day. Only a tight-weave, lined wind-breaker kept Colleen warm. The fishermen seemed immune to the cold weather, but many residents over the years moved their homes back from the coast, into the foothills, to get more sun.

Now that she could afford it, Colleen found herself buying clothes more often, paying more attention to the style and fit. She laughed at her own vanity until Jonah teased her, and then she fought back.

"I wasn't born vain," she told him. "I learned it. I feel social pressure to dress stylishly, especially when I have to talk to students, and faculty, and big shots who I'm supposed to impress. Now I'm appearing on national TV. They'll judge me by my clothes, and you know it, smart ass." The speech ended with an anger that even startled her.

"I wasn't criticizing," Jonah said, abandoning the field. "You look great. You're prettier every day."

Now that Colleen dressed more carefully, she didn't jog to school. As she tried on her new tailored suits and dressy skirts and blouses, she appraised herself. No avoiding it, she thought. I prefer the new look. But I have to get more exercise.

She rode a bicycle to the campus, hurrying to approve a BBS

broadcast script. When she arrived, she went to a small studio, watched the video, and edited the script, making frequent changes. The script was not finished and ended in mid-sentence.

The studio door opened, and a young man stepped in with a sheaf of papers in his hand. Colleen stood up, and the newcomer stopped to stare. He looked Colleen over, openly admiring her.

Finally he said, "Excuse me. I'm supposed to take the rest of this script to Dr. Jameson." He started to back out of the door, still looking at her.

"Here, let me have it. I'm Colleen Jameson." So you want him to call you Colleen? Why did you work so hard for that damned degree if you don't want him to call you "Doctor?" She examined him in her confusion. He stood over six feet tall, muscular, well-built, and tanned. He looked very powerful. Colleen felt warm and took off her sweater.

"Have a seat," Colleen said. He crowded into the sound-proof cubicle and sat in the only other chair. Knee-to-knee, near the TV monitor as they talked, she wanted to know, "How do you keep in shape? Do you swim?"

"I did for the last two years, while I lived in Hawaii. An Olympic pool, every day."

"Hawaii, and not in the ocean?" Colleen asked.

"Ha," he said. "Better not. Hilo Bay is polluted with sewage and has arsenic all over the bottom."

"You're kidding."

"Unfortunately, no. Hawaii is one of the most environmentally spoiled places you'll ever weep over." He stopped again. "You're really Dr. Jameson? By the way, I'm Evan. Evan Thornton." He took her hand in both of his.

He gave Colleen the sheaf of papers, and said, "Sorry, I just finished this."

"The rest of the script?" Colleen asked.

"Yeah," he said. "I asked outside if I could work on newscast scripts. They said they'd try me out."

"You wrote this script?" Colleen asked.

"Yeah," Evan said. "What does it need?"

"I made a few changes. What's your background?"

"Degree in English," he said. "Worked two years as a reporter for a small paper, the *Hilo Tribunal*."

"Well, if you have time, sit in on a beginning biology class. You should look up the problem of anthropomorphism and teleology in science writing."

"Any clues?" Evan asked.

"Well, it's better not to assign human motives to other organisms. And don't invent things that aren't there, like 'the balance of nature.'"

"Really? There's no balance in food chains?"

"Not really. Just a bunch of plants and animals eating each other." She gave him the edited script.

"Well, I'll go make the changes and see what I can learn." He moved out of the door with athletic grace, then turned back. "You're really a model waiting for the videotape crew, aren't you?" he asked, grinning.

Colleen laughed. "Go fix the script."

She returned to the videotape but couldn't concentrate.

47

BAT DEBT

Bat Masterson's cheerful face smiled out at the evening viewers of the BBS "Jameson hour" newscast.

"Good evening, ladies and gentlemen. Tonight, I have to give special thanks to the Burner Broadcast System. I went to the major broadcast corporations, as your nation's acting Secretary of the Treasury, and asked them to give me air time to bring you a very important message. All I can tell you is that they tried to censor the program I prepared. For that reason, you might want to take a moment to call a special friend to watch the next half-hour. It will tell you something that the major networks don't want you to hear."

Records later indicated that the telephone network across the nation showed a large jump in traffic. The special attraction of the BBS newscasts came from people expecting to hear the inside story. For the rest of the hour, Bat had the highest number of viewers of any politician seen on national TV.

"Some of you have probably heard me talk about the people of this great country as a big family, not exactly like many of our small families, of course, but a group of people who need to feel the benefits of being in a family."

"I like to hope that living together as a family provides care for the smaller, weaker ones: some guidance and support, or just a friendly ear when one is troubled; a place to rest when weary; a place to express your distress, even anger, if frustration overwhelms you; people to love and worry about so you feel part of the human race."

"My liberal friends say I oughtn't use the family image to scold parents who live apart, and I don't mean to do that." Bat shook his head in dismay. "Most families break up sometime. I'm also aware that families can be a dreadful place where weaker members are mistreated."

"However, I use the family notion to refer to people who support those who live near them. And we all should do that. You all know my background, and you'll forgive me, I hope, for this little sermon."

"But what has this to do with my job as Secretary of the Treasury? Let me tell you."

"Recently I went to the library and got a copy of a very well respected college textbook on economics. I reviewed for myself the current thinking on the role of the federal debt, those five thousand billion dollars we owe ourselves. Most of our legislators think this is an unhealthy situation, and they have tried to reduce the national debt, often by raising taxes, and trying to keep the federal budget down. But the debt has climbed steadily, for both conservative and liberal administrations and congresses over many decades."

"Now, I'd like to tell you about the Bat Masterson Family Plan for debt reduction. In the economics textbook I mentioned, it seems that a popular way to explain the national or internal debt is to imagine that family members lend each other money."

"My grand children agreed to help me with this little story," Bat said. "For example, let's imagine a growing family with two older sisters who have jobs." Here, Bat waited while two teen age girls, one white, and one black, walked cheerfully into view and faced the camera. Bat then waved a ten year-old boy on stage.

"This is Johnny, a younger brother who needs a bicycle but has no money." A boy walked on stage and stood on one foot.

"Now suppose the parents thought that Johnny needed a bicycle more than the girls needed their money. Perhaps Johnny could get a job with the bicycle. So, the parents, serving as the government, said, 'We will issue a $100 bond and one of you girls who has some money can buy it, to be repaid with 10% interest in one year, $110 in all." Bat handed one of the girls a large green sheet labeled 'Bond,' and the girl gave him $100.

"As the government, I can decide that Johnny needs this $100 for a bike so he can take a job as a delivery boy." Bat handed the money to the boy who rushed out and wheeled a new bike into view. Bat continued.

"Johnny got his bicycle, the family internal debt became $100, and in a year the bond will come due. To pay off the first bond, the parents issued another one, this time for $110, bought by the second sister." Bat handed another green document marked "$110" to the second girl. She gave him eleven $10 dollar bills.

"This $110 paid off the first bond which came due." Bat transferred the money to the first girl who smiled at receiving her $10 interest.

"What has changed? The family debt has now grown to $110, owed to each other."

"Now watch closely to see how the Bat Masterson Family Plan reduces the national debt."

"Suppose that the second sister who owns the $110 bond is about to get married and leave home." Bat stepped to one side, while the second girl faced Johnny, and said in a clear voice:

"I'm leaving, Johnny, and I don't need the money as much as you do. I'll forgive the debt." With this, she tears the bond in pieces and discards it.

Bat then said to her, "That's very good of you."

Bat turned to the TV audience. "With my background, I would call this the Christian thing to do." On cue, the young people filed off, with the black girl giving Bat a quick kiss as they left. Bat watched fondly as they departed, finally turning back to the audience. His demeanor became more serious.

"But how does this apply to the real federal debt? Here is my family plan. Our corporations hold a large number of the bonds which make up the national debt. Corporations aren't like people, not real flesh and blood. So we will take back the bonds owned by these corporations, and return them to the government as a debt that has been forgiven. This way, we can reduce the federal debt down to nothing. In addition, we allow the business community a way to participate in the real charity of Christianity in this country, in a deep and meaningful manner."

"In the event that our corporate leaders are shy about stepping forth to participate in this program, I worked out a procedure by which we in the Treasury Department will select the first corporations, until all of them fall into the charitable spirit."

"You probably heard a list of corporation names read during an earlier broadcast. Well, recently, I have learned how these corporations have been less than friendly to our human family, causing great damage to the world which supports us all. So we will start by defaulting on the bonds held by those corporations. We invite the Jameson Population Institute to nominate more corporations which have been the most selfish family members. We at the Treasury Department offer our assistance to help bring them back into the family fold."

"For those of you who think this is harsh, let me remind you that last year alone, American businesses took in 15 trillion dollars in receipts. Our total national debt is not even a third of that amount."

"Now would you please pray with me?"

Bat Masterson put his head down in his hand and in his clear resonant voice called out, "Oh Lord help this great American family take good care of each other. Help make enough to go around. Help our viewers call these numbers and show their support. Amen and good night."

Bat smiled and waved as the newscast closed showing Washington DC telephone numbers for the President's office, the Speaker of the House and Senate, and many other decision leaders in congress. Viewers blanketed the phone numbers for weeks.

48

. . . old people have been brainwashed and pacified into believing they are powerless . . .[and] insulted by labels referring to old age as "second childhood." . . . This has truly terrible consequences. Many of the old hate themselves and their age group, seeing nothing useful or interesting in late life. They believe they are decrepit and burdensome to their families and society. They have no appropriate outlets for aggression and anger . . .

From *Why Survive? Being Old in America,* RN Butler.

Unexpressed emotion tends to become pathological.

Psychotherapeutic dictum.

TERMINALS

Earl parked his car, shuffled into the lobby of the patients' treatment quarters, and walked toward his room.

"Hey," a voice called as he passed an open door.

"Who, me?" Earl inclined his bald head around the door jamb.

"I'm Arvin." The white-haired man pulled a clean shirt and a bottle of vodka from a suitcase on the bed.

"Whatcha in for?" Arvin blurted out, peering through thick spectacles.

They examined each other, looking older than their 70 years, bent, in pain, and unnaturally thin. Earl wanted to give up the treatments but was afraid to go home and just sit and die.

Earl said, "Cancer, radiation and chemotherapy."

"Uh oh," Arvin said. "Me too. What's yer name?"

"Earl." They shook hands.

"It's gettin' dark," Arvin said. "Before my wife died, she wouldn't let me drink 'til then. Want a belt?"

"My doctor said 'No booze.'"

"So did mine," Arvin said. "But I think I'm terminal."

"Me too," Earl said.

"So screw the doctors and sit down." Arvin poured vodka in two glasses and broke ice out of a tray. Earl brought a chair and a can of peanuts from his room. They talked a bit and aired out the smell of death in the room.

They began to visit every afternoon at the cocktail hour pretending they were in a bar out in the real world.

"Didja watch Public TV today?" Arvin talked about the programs he watched and waved his arms for emphasis. "The doctor that showed how cancer rates shot up since 1950?"

"Why since 1950?" Earl liked Arvin's stories.

"That's when they started using agriculture chemicals on crops. How come we never heard about that before?"

"People would have sued them." Earl nodded his bald head to punctuate his comments.

They criticized the newspapers, the magazines, and TV for ignoring their health problems and their fixed income in the face of inflation.

"Well, they better not cut our social security payments!" Earl liked to sound a bit rowdy.

"They'll try. Hey, it's time." Arvin flipped on the small black and white TV screen to their favorite newscast.

The announcer began. "We bring you a special presentation of scenes of the recent violence during the Florida county council meeting, recorded by a cable access crew. We urge viewer's discretion," the announcer said piously.

Earl threw him a finger. "What? I should close my eyes?"

The TV screen showed the lawyer and the suit-and-tie businessman earnestly exhorting the county council with a few plainly dressed people listening.

Earl and Arvin then sat and watched Patrick Stevers, the old man in overalls, approach the lawyer and businessman. He waived his fist and started shouting. After telling how he lost is land, he shouted:

"That isn't fair."

Earl and Arvin hung on every word and watched Stevers shoot up the council room.

When the program ended, Earl snapped his mouth shut. "Geez. I forgot to drink my wine."

"That was the damnedest thing I ever saw," Arvin agreed.

The next night, Arvin immediately started talking about Patrick Stevers.

"I listened all day to what the commercial networks said about that guy who shot up the County Council. Did he look like a cowering fugitive, holding hostages?" Arvin asked.

"Not to me," Earl said.

"Did he look crazy, like he didn't know what he was doing?" Arvin asked, waving his arms.

"No sir," Earl said. "I heard that bullshit from the commercial commentators, too. He was calm, almost peaceful."

"Right. Did he sound like a political eccentric with an axe to grind?"

"Nope." Earl shook his bald head. "He didn't bitch about liberals at all."

"So, what did we really see in Patrick?" Arvin asked, getting angrier as they talked.

"Well," Earl said, "he worked hard for a living."

"You bet. Then what?"

Earl responded with relish. "Somebody took away his livelihood."

"And he refused to accept what they did to him," Arvin said.

"Right," Earl said. "He fought those bastards right out where people could see him. What's so bad about that?"

"Was he fearful? Cowardly?" Arvin asked.

"Angry," Earl said. "Proud, stubborn, and angry."

"Now," Arvin spread his arms dramatically, "could you do what he did next? He swallowed a lot of pills and drank a lot of whiskey."

"I usually drink the whiskey and take the pills the next day," Earl said. "But I've been doing that for years."

"To get what he wanted, did he worry about the police clubbing or shooting him?"

"Nope," Earl said. "He made sure that he passed out before they could get to him."

"And now, the best of all," Arvin said. "Did he look selfish? Did he do bad?"

"Hell no," Earl said. "He died a hero, as far as I'm concerned."

Exhausted of righteous indignation, their conversation slowly wandered back to their regular fare of bitter complaints against the establishment, the goddam wealthy people and their banks and lawyers, the stock market scams, the crooked schemes to keep the unions powerless and the working people poor.

The next night, Arvin asked, "You ever feel cheated like that Patrick fellow?"

"All the time," Earl said. "I bought a car that broke down, and spent $15,000, and they never did fix it. They all put me off, the salesman, the dealer, the repair garage, the Better Business Bureau. Even the State Consumer Advocate. I wrote the manufacturer. They all lied to me and told me they couldn't help. I could'a killed 'em."

"Is that right?" Arvin asked.

"Is what right?" Earl asked.

"You said you could have killed them. Why didn't you? Were you mad enough?"

"Yes. I still am, whenever I think about it," Earl said.

"Would you like to kill 'em now?" Arvin asked.

"What? You mean really kill them? Like shoot 'em?"

"That's what I mean," Arvin said.

"Well, yeah. I think so. But I don't want to go to jail," Earl said.

"Patrick Stevers didn't go to jail," Arvin said.

"Yeah, but he died."

"He died after he had done something that he really wanted to do," Arvin said. "And he died right after he chose to drink a lot of good liquor and pass out."

Earl stopped chewing on his fish and stared at Arvin.

"What will you be doing when you die?" Arvin demanded. "And when are you going to die? And how?"

"Jesus," Earl said. "I'll prob'ly be full of pain pills, still hurting like hell, full of needles, avoiding the horrified eyes of my daughter. She'll see a weak, beaten old man she hardly knows, clinging to a life full of fear and agony."

Earl sat stunned, thinking about the feelings which had just poured out of him. How simple the world is, he thought, when you clear out the bullshit.

"That's about how I see it," Arvin said. "My doctor said I have one to three months. Do I want to die looking like a pincushion? I guess I'm lucky I don't have a daughter to watch me die."

Earl and Arvin talked much more seriously after that night. They talked about who they really thought was responsible for people's problems.

"The government?" they wondered.

"Nah," Earl said. "Bunch of deballed bureaucrats defending their territory."

"Television?"

"Definitely, TV has caused us old people problems. They keep comparing us to young people," Arvin said.

"Why don't we think it's the government?" Earl asked. "Everybody else thinks the politicians are the problem."

"Yeah, but if you shoot one, another just like him pops up," Arvin thundered.

"But that's because the same businessmen buy their votes," Earl said.

"Well, then, we should go after the businessmen," Arvin said.

"Just any of them? All of them?"

"The bigger the better," Arvin said.

They wondered if their illness had made them a bit loony, so they tried an experiment. They called old friends and asked a question.

"Is there anything you would really like to do," Earl asked his friends, "but won't because the police would lock you up for good? Anything?"

The first answered: "I had a boss I hated so bad I used to wish someday I would catch him near the edge of a cliff. I'd have pushed him over for sure."

The next told them, "There's a lawyer and an editor in town that I would take pleasure in shutting up forever. They are really mean."

A man said, "I may yet beat my boss to death with the hammer in my desk drawer."

A woman told them, "Once in a big supermart, I could have turned and walked away from my kids and husband without anyone seeing me go. I wanted to do it. I almost did it, and I still wonder what would have happened if I had."

Earl and Arvin decided their own feelings weren't so crazy.

Earl brought around an invitation to a stockholders meeting for a big auto company.

"Where's the meeting?" Arvin asked.

The board meeting would convene in a luxury hotel near the international airport. Earl and Arvin plotted furiously and came up with a plan they hoped they could carry out.

Then one night at the cocktail hour, Earl faltered.

"Now that I've looked at our plan up close," Earl said, "I don't think I can go through with it. I'm not sure I can throw the match to start the fire. A lotta people will burn to death."

"That's true, no doubt about that," Arvin said. "By the way. Remember the car manufacturer that put out a vehicle with a dangerous gas tank?"

"Sure," Earl said.

"Then the safety organization showed that the manufacturer knew fires would kill people in some accidents," Arvin said. "They proved that the company CEO delayed complying with federal safety regulations for eight years."

"Yeah, I remember," Earl said.

"Eye-for-an-eye," Arvin said. "They burned people for eight years. We burn them."

"But I'm not a Christian," Earl wailed.

"No, but read this." Arvin gave him a newspaper clipping.

"Detroit Motors stalled putting safe gas tanks in their car by presenting a Cost/Benefit Analysis. They showed Congress that in eight years safe gas tanks would cost $1.1 billion to prevent $400 million dollars worth of deaths, injuries and damages. The manufacturer decided: *Let 'em burn.*"

"Jeez," Earl said. "What bastards. I still don't know if I could do it, though."

"OK by me," Arvin said. "I didn't lose $15,000 on a lemon. Let bygones be bygones. Turn the other cheek, I say."

Earl nodded.

"However," Arvin said, "I may do it by myself."

Earl frowned.

They turned the conversation back to usual topics, and Earl lasted three nights.

"I been thinking about what kind of people would let their customers burn to death just to save money."

"Yeah?" Arvin asked.

"Those executives say they have to show quarterly profits or the Board of Directors will fire them."

"Yeah," Arvin said.

"Well, both the board and the CEO say they'd rather burn people to death than lose money. Right?"

"That's right," Arvin said. "You could say that both the CEOs and the Board members are vicious inhuman executioners willing to kill painfully for money."

Now Earl sprang his trap. "So why shouldn't we kill them to prevent this kind of cruelty in the future?"

"By God, Earl, you've overwhelmed me with your logic."

Earl happily returned to the fray.

Earl and Arvin moved into the hotel where the automobile manufacturer scheduled the meeting. Amidst the plush accommodations, they brought in suitcases concealing two wagons, in kit form, and gallon cans of paint thinner. They tried to assemble the wagons, and nearly gave up the project because lever A didn't always fit in slot B. Finally they loaded the wagons with open cans of paint thinner, gasoline, kerosene, and acetone and taped everything in place.

They reconnoitered and found the luxuriously draped conference room with the Chief Executive Officer showing profit and growth curves that went up, and up, and yet farther up. Back in their room, Earl and Arvin put on painter's garb and threw drop cloths over each wagon. With paint brushes on the drop cloths, they trundled off to the conference room with its two exit doors.

Each pushed his wagon inside a door, and dumped it on its side. The stockholders saw the gushing cans sliding toward them, prominently marked "Flammable," an evil mixture of volatile fluids. When Earl and Arvin fired marine flare pistols into the pile of gurgling cans, the acetone and gasoline exploded in a fireball. The kerosene and thinner boiled in the cans, gushing out spurts which burned more determinedly, an expanding, smoking sheet of flame, moving in toward the occupants of the room. They retreated, screaming, into the farthest but windowless corner.

"Fire, fire," shouted Earl and Arvin. They hurried down the fire escape stairwell, stripping off the painters' clothing. They walked to their car and slowly drove back to the patient quarters. They watched television all day, bone weary and dozing from their exertions.

Most of the people in the room died. Earl and Arvin watched, awed and sobered, but not quite remorseful. They clutched sleeping pills and vodka in case the police started beating on the door.

When the authorities were unable to trace them, Earl and Arvin felt a twinge of disappointment, and compensated by writing a long letter to the editor. They explained about the need for patriots like Patrick to defend the public against ruthless white-collar criminals who ran corporations without regard for human well-being. It was a stirring letter, and Arvin had to restrain Earl from signing it.

None of the editors printed it. Then Colleen's investigative reporters got hold of a copy, and read the letter on the Burner Newscast, with a disclaimer.

"Without endorsing reactions this strong, we find the reasoning persuasive. The dead stockholders had caused car drivers to be burned and maimed with inexcusable, even monstrous lack of regard for human welfare, in order to secure financial profit. We acknowledge the rage of the people who struck at their oppressors." The newscasters ran old films of the cars burning.

The Burner Broadcast then gave the names of the editors that refused to print Earl and Arvin's letter, which in effect censored the letter writers and protected the corporation from criticism. The BBS called this level of partisanship "profoundly anti-social, even treacherous in a democracy which so desperately depends upon the truth."

Earl and Arvin's exploits received constant attention. They were overwhelmed by their feelings as they listened to commentators criticize and praise them.

In the weeks that followed, Earl often thanked Arvin for removing fear from his life.

Arvin's doctor talked about some improvement in his condition, "especially, your morale."

"I've been getting some exercise," Arvin told him. Exercising my body, my mind, my problem-solving ability, my autonomy, my self-expression, my rage. He told Earl about his feelings. "I've been defending myself. I've given up being a victim and feeling helpless." He laughed. "I had to get terminally ill to learn how to live, to stand up for myself."

One day Earl asked Arvin, "Did you know that there's a directory of those people that sit in our legislators' laps, stuffing money in their pockets, and whispering in their ears?"

"Lemme see," Arvin said. "You got a directory of lobbyists?"

49

ANIMAL

Today, Colleen put on a new skirt. She drove her car to the campus, and wondered, Why am I driving, not jogging? She couldn't stop herself from wondering whether Evan would be at the Media Center, working on the Burner newscast material.

He sat punching a computer keyboard rapidly.

"Hi." She tried to sound businesslike.

He turned with a look of irritation then broke into a smile. He jumped up to pull a chair near his.

"God, this is great. I was just thinking about you." He stopped for a moment. "You look marvelous."

He barely touched her arm to escort her into the chair as she sat down. He turned to the video screen and said, "I wondered if this script is better. Would you look at it?"

She leaned forward and read. She liked it better than the material from the previous day. "I like your description here. It doesn't editorialize, or draw the conclusion for the viewer. I take it the video will show the marshes that the developers plan to fill?"

"That's right," he said, pleased.

"Now, this part. You mention losing the beauty of the marsh and the birds which fly through. I agree that it is beautiful, but we also need to point out that the marsh produces the biomass which nourishes this entire area, including the fish offshore."

Evan turned toward her, watching her eyes and mouth. He lost his focus and couldn't follow her words. "So scientists don't attend to beauty?"

She saw that he stared helplessly in her eyes again. He forgot the writing. He leaned close to the screen and she could smell his body. She shivered.

"You asked about swimming the other day," Evan said, quietly. "Would you like to do a few laps in the University pool and then maybe have a sandwich? You could tell me more about the writing?"

"Uh, I suppose so," Colleen said.

"In an hour, then?" As he stood and moved away he lightly touched her arm. She jerked involuntarily.

She watched him go, wondering, What am I doing? Then she leaped to her feet, ran out to her car, and drove to the closest dress shop. She tried on bathing suits and bought three.

She went to the pool dressing room and put on the most revealing suit. She showered, and as she stepped out of the ladies dressing room, she saw him waiting. She tried not to stare at his slender waist and hips.

"Shall we?" He led her to the lap area, and they jumped in. "Have you done much swimming?"

"Not recently," she said, shivering.

"Better make it just a few laps for now."

She pushed off, and he swam along beside her. After a few laps, she stopped for lack of breath in the deep end.

"C'mon." He took her by the arm and pulled her a few yards to the edge of the pool. They clung there as she panted. He placed his hand under her arm and supported her. "Are you all right?" he asked.

She nodded, aware of his hand against her breast and reached toward his waist. She closed her eyes and breathed deeply, letting herself drift with the feel of the water, his chest, his arm, and his waist. She pulled herself toward him until their bodies touched.

"Will you meet me outside?" he asked. "I'll wait in the parking lot."

She nodded and they scrambled out of the pool. When dressed

she walked out of the gymnasium and saw him standing by a camper. They got in, and he turned to look at her. She looked like a bedraggled puppy. He took a towel from behind the seat and dried her hair. He found a brush and stroked it until her eyes closed. She leaned against him and held his thigh in her hands, rubbing the hair on his leg.

He opened the center door to the camper, slid himself into the back, and pulled her after him, onto a wide bunk. He shut the door, and they were closed away in the dark confines of the camper. She could smell his body again.

He lay down beside her, put his arms around her, and shivered violently for a few seconds. Then he slowly began to kiss her face and lips. He slowly undressed her and himself. The strength of his hands excited her as he moved them about on her back.

She breathed hard, writhing, wanting him to touch her all over. But he stayed with her face and breasts, enjoying her with his mouth and hands.

She reached down and held his penis, squeezing and feeling the strength of it. She moaned and tried to pull him between her thighs. He kissed her on the stomach and licked and bit her gently, slowly moving his mouth lower.

Colleen's body became so taut she thought she would burst. He kissed and licked her for what seemed an endless time. She moved her hips against his mouth and finally came, waves of convulsions racking her.

She panted and gasped as she recovered. Then she took his penis in her hand, moved herself over him and guided herself down around him. She moved slowly as he tried to hold back. Finally he rolled them both over. "Kiss me while I come inside you," he whispered, urgently.

She nodded, locked her arms around his neck and her legs around his waist. Then they let themselves go.

That night, Colleen lay awake in bed beside Jonah, trying to feel guilty. But she only felt glad to be there. She thought about all the times he had helped her and protected her. She knew the feel of him, his smell, and his taste, and she felt his warmth. She gently pressed herself against him and felt safe.

Jonah rolled over. "You're awake?"

"Yes," she said. "How did you know?"

"I could hear your brain whirring. You all right?" he asked. He put his hand on her face and kissed her.

"I am," she said. She turned her face and kissed him. She thought back to the first night she met him. She remembered how he gently wiped her with a warm damp washcloth, and made her feel safe enough to sleep.

"Yes, I certainly am," she said. "Why wouldn't I be, with you here?" She wanted to give herself to him, if he wanted her now. She kissed him again, gently, without demanding.

Jonah kissed her back, slowly becoming aroused. "You never hold back from me. Do you always enjoy it?" he asked.

Colleen scarcely heard him. She was still inflamed from her encounter with Evan. Her skin tingled and swelled under his touch. She became mesmerized with her own body and her reaction to Jonah's. She kissed him ever so slowly, and murmured encouragement.

Jonah breathed more deeply as she kissed him. He whimpered when he couldn't hold himself back. Colleen slipped him inside her and lay perfectly still while she kissed him. Her complete acceptance of him drove him wild, and he had a shattering orgasm without thrusting in her. She held him as her body accepted his shocks.

At the last moment of his orgasm, she caught fire, lunged back and forth, and exploded with her own.

She held him for a long time, and kissed him back into a peaceful place in her mind. Then she floated into a multi-colored sexual swirl of pleasure and drifted off to sleep with Jonah still inside her. Her last thought was, "I'm an animal."

50

I sympathize, therefore, with those who would minimize, rather than with those who would maximize, economic entanglement between nations. Ideas, knowledge, art, hospitality, travel—these are the things which should of their nature be international. But let goods be homespun whenever it is reasonably and conveniently possible, and above all, let finance be primarily national.

J.M. Keynes

FOUR TRILLION DOLLARS?

Soon after the President of the United States appointed Bat Masterson to be Secretary of the Treasury, Bat fulfilled his threat to default on certain notes and bonds making up the national debt. Repercussions spread around the world, and worked their way up to the Director of the World Bank.

At the moment, in an expensive suite in Beverly Hills, the Director awaited the arrival of five CEOs of large multinational corporations. He wished himself back in New York in his old job as Director of the International U.S. Bank. He hated Los Angeles. After the delay through customs and baggage collection, he had navigated the carnival of vehicles in search of his limousine. Los Angeles was not yet as bad as Lagos, but bad enough. The Chairman of the largest Italian multinational corporation had called him in Djakarta.

"Hello, my friend, how are you? I would like to come see you next week in Los Angeles. OK?"

"No, it is not OK, " complained the Director.

The Italian ignored him. "Our friends in London and The Hague set it up. This time they invited us. Usually they think they're running everything all by themselves."

"That's true," the Director said. "What's up?"

"Imagine those guys, bashful," the Italian said. "They wouldn't give me a clue, but I assume it is about a fellow American of yours."

The Director smiled at the colloquial flavor of his friend's English. Must be his new mistress. "Please tell Veronica she is having a good effect on your accent. You now sound like a Hollywood starlet."

"Don't tell me that," laughed the Italian. "My tongue will freeze in my mouth if I think of that. I will bring her and you can tell her yourself. You coming?"

The Director concluded that they had Bat Masterson problems. "I wouldn't miss it," he said.

Problems for the Director of the World Bank started over a decade ago, before he became Director. He was one of the leaders of the international banks who urged countries like Argentina, Brazil, Mexico, and Nigeria to go on the biggest borrowing spree in history. The World Bank had secured from its donors very large sums, and increased its lending ten fold. MacNamara funded projects he said were "to reduce poverty," his cover story while he directed the World Bank.

When these major borrowers threatened to default, the US wanted the World Bank to press them to reduce imports and increase exports to pay the debt service to the World Bank. Forcing them to export reduced those countries' sovereignty and turned them into export machines, which of course took food out of the mouths of the poor and provided cheap raw material for the Northern hemisphere.

That's when the Director joined the World Bank. The press characterized his policies as squeezing money and goods out of poor people. Screw'em, he thought.

But then the environmentalists got Congress's ear. God, those endless inquiries. The Director claimed the Bank always did careful environmental studies before funding any project. But that wasn't true, and they proved it. Everyone knew that he, and MacNamara before him, had lied.

The US Congress then heard of the World Bank's Plan 2010 for Brazil to provide 136 dams, 80 in the rainforest, which would destroy the Amazon Basin and several Indian tribes. All of the dams would be unnecessary if, instead, they invested in end-use energy efficiency and conservation.

Now he was in Beverly Hills, and the World Bank Director waited to referee a decision by the CEOs of the largest transnational corporations in the world.

After they arrived and settled in, they talked most of the afternoon about the effects of Bat Masterson's refusal to honor a large quantity of bonds. They had only lost small amounts of money so far, but decided they could lose a great deal more money if Masterson continued his default policy.

"Why is he doing this?" the Italian asked, surprised at the bravery of a man who would say and do these things in public.

"Well," the Director said, "we could have predicted it after he ran for President. He ran as a television evangelist, a fundamentalist Christian, a Republican for Christ's sake, and he promised on TV to forgive the national debt if elected. He ran eight years ago and did not do well, so everyone forgot about him."

The Hague representative said, "Forgive the debt? Four trillion dollars?"

"He claims it's the Christian thing to do," the Director said.

"But he lost the election," the British CEO said.

"Yes, but our goddam liberal President, in his wisdom, appointed Masterson Secretary of the Treasury," the Director said. "Of course he did it to punish his Republican opponents."

"Four trillion dollars?" repeated the CEO from The Hague. "This President of yours is capable of almost anything."

"I'm afraid so," the Director said.

"Four trillion dollars? He has to be stopped. Completely. Soon." The Hague CEO suddenly shook himself, and looked around at his colleagues. "You agree of course?" They did not. But what The Dutchman wanted, The Dutchman got. The American President had to be stopped.

The World Bank Director got the job of making arrangements to "stop" the President. The others believed that the Director's old US International Bank had Mafia connections. These executives would buy help wherever they could get it and would be pleased to use reliable craftsmen with a history of success. Besides, the Mafia might be angry at this President, too.

The Hague CEO mused to himself. He'd gone too far, this American. Oppose us, would he? he said to himself in Dutch. We'll show him. We didn't set up NAFTA and GATT to avoid all those trade barriers so we could make a lot of money and then let some short-term politician try to stop us. An American, yet! It just shows how poorly educated they are. They really don't understand how the world is run.

Soon they arranged a very large contract against the goddam liberal President's life.

51

CITIZEN UNREST

Colleen's broadcasts on Burner Broadcasting System continued full force, and many credited them for stirring up the citizenry. During the weeks after the first fire-bombing of a corporate board meeting, citizens complained against the government ever more bitterly. An angry group of middle- and working-class women joined the crowd of senior citizens permanently camped outside the White House. TV showed frustrated policemen trying to restrain old women clutching feebly to their walkers.

Muslim and Christian blacks appeared, shouting on soap boxes amidst the crowd. As each day wore on, the crowds sang, chanted, and shook their fists at the White House. They threw trash and garbage over the fence, onto the White House grounds. Guards inside first scolded the throwers and then became targets. Part of the crowd stayed each night and used the portable toilets brought by construction unions. By the second month, thousands of people camped everywhere. They used blankets, plastic sheeting, newspaper, and cardboard for sleeping.

The President prayed for rain and resisted the angry businessmen's demands that he drive off the growing encampment of angry people around the White House. His aides new better, remembering White House history.

The crowd displayed signs demanding no compromise. They shouted that no private money should be spent to elect public office holders. They didn't want private or public tax dollars spent advertising any product or service. No one knew, yet, how to write that law. The old "truth-in-advertisement" law certainly failed.

The sun-belt provided the most demonstrations, in Miami, Phoenix, and Southern California. There, the state legislatures capitulated early and passed versions of the Free Elections and No Advertising laws. Demonstrators in those states congratulated their own politicians and demanded that the federal politicians follow suit. The battle cry then became, "Free Congress from campaign contributors," and "Free TV from advertisers' control," or simply: "Free Congress, Free TV."

Leftists taunted business groups, saying it was time they lost ownership of the media and its thought-control over everything Americans did.

On the day after her encounter with Evan, Colleen hurried from the Population Control Institute to the University Media Center to provide an introduction to a videotape for the BBS newscasts. The videotape had been prepared at the University of Oregon where faculty members made estimates of the increase in quantities, chemical species, and radioactive decay progress of the nuclides leaking from Hanford, Washington into the Columbia River. These wastes currently reached dangerous concentrations.

Starting in World War II, the Atomic Energy Commission operated nuclear reactors at Hanford for the single purpose of transmuting uranium to plutonium for weapons. Not even used for electrical generation, the heat produced destroyed life forms in the Columbia River.

Videotape producers frequently asked Colleen to introduce programs that appeared on BBS newscasts. It was clear that Colleen had polarized the media when a news commentator quipped that there should be no more "broad" casts. Feminist columnists and women's organizations got him fired.

When Colleen entered the media building, she saw Evan scolding a typist.

"You think you're good enough to edit my stuff? Don't ever change anything again?"

Colleen asked, "What's wrong?"

Evan turned to see her and smiled radiantly.

"Nothing." Turning to the operator who sat stunned and resentful, Evan said, "Don't worry about it. I can always change it back, right?" He chuckled and took Colleen's arm, walked her to the studio, and handed her the videotape script. "What time should I pick you up tonight?" he asked.

"No. Oh no. We have to talk. I don't want to . . ."

Evan smiled, put his finger to his lips and interrupted her. "Shh, not now." Still smiling cheerfully, he gently pushed her into the studio, and closed the door. Colleen watched him walk away. What in god's name was he thinking?

"I'll have the tape ready in two minutes," the engineer said over the intercom. "Are you ready?"

She nodded and turned to the script which Evan gave her. She found herself frantically editing to restate the message in a more palatable way. She winced each time she came to a personalized and accusative comment. She found herself thinking, Well, he started as a reporter. He's been rewarded for writing exaggerated descriptions throughout his career. But his tone and lack of logic almost suggest he's trying to get me into trouble.

When she finished the videotaping, she went looking for Evan. The negative image his scripts conveyed disturbed her. She saw him talking to a woman who reached in her purse and handed him something. With a severe look on his face, Evan nodded, turned and headed out the door. Colleen followed, determined to talk to him.

On the sidewalk outside the building, two men and a woman, all in suits, bore down on Colleen flashing badges.

"You are Colleen Jameson?" the woman asked. "Please come with us."

Colleen turned on her heel and walked rapidly back into the door of the Media Center. She shrugged off a hand on her shoulder. Just into the building, she turned and blocked access to the door.

"What do you want?" she asked. "Who are you? Keep your hands off me," she said loudly.

"We are agents of the Federal Bureau of Investigation. We have orders to take you with us." One reached to take her arm.

"Keep away. Don't touch me," she shouted loudly. Several students turned and hurried up to Colleen. "What's the matter, Dr. Jameson?" one asked.

"These people are trying to kidnap me," Colleen said. A crowd grew in and outside the building.

"How do I know you are FBI? Let me see your identification," demanded Colleen. She took their badges, looked skeptically at them, and handed them behind her to friends on the video team. "Somebody call and see if these are fake," she said loudly. A young male took them and gleefully headed for a telephone.

With the help of her friends, Colleen closed the bottom half of the Dutch door and bolted it shut. Others ran around and locked the rest of the entry doors. Colleen continued to shout at the FBI agents, telling them to go away and leave her alone. Dammit, she thought, I said I wouldn't run anymore, and I won't. By now several dozen students surrounded the building. A few took up a chant: "Feds off campus," but the FBI agents stayed at the door.

The Media Center video operators trundled camera tripods and light sources to where they could catch the action in and out of the building. All attention focused on the FBI agents who just wanted to get their man—this time, a woman doctor. A crowd of students now surrounded the FBI agents.

The President of the University arrived, carrying a loudspeaker, in the company of the County Sheriff. The President announced that since no one warned him of the arrival of federal officers, all unauthorized persons should leave the campus immediately. The students roared their approval.

Encouraged by their President's comments, the students began to taunt the federal agents and push them away from the building. One male student who found himself next to the female FBI agent loosened his fly and cheerfully urinated on her.

She flew into a rage and pounded him about the head: "You . . . little . . . bastard. How . . . dare . . . you?" The other students jeered at the beleaguered agents.

A car arrived and unloaded a half-dozen more FBI agents who started toward the crowd. The college President again told them they trespassed. The Sheriff and a few of his men tried to interpose themselves between the new group of advancing federal agents and the crowd of students.

Several male students near the female agent began to grope her. One held up her cellular phone, displaying it as a trophy. Nearby students urged on those close to the female agent.

"Get her gun. Get her bra." All three agents now gripped their weapons to prevent losing them to the grasping hands of the students. The new group of FBI agents tried to reach their colleagues in the midst of the students, while Sheriff deputies pushed them away. The President implored the group to break up. Finally, the inevitable happened.

One of the FBI agents clutched his gun while a nearby student tried to wrestle it from him. The gun fired, hitting a female student. Those nearby jumped back, leaving the agent in a circle of students, the wounded student writhing on the ground in front of him, his gun pointed at her.

The footage taken by the television cameras did not clearly display the student provocations. However, it did show the College President demanding that the FBI leave, and it showed the Sheriff deputies trying to restrain the FBI agents.

The crowd of students grew to several hundred while an ambulance removed the wounded girl. Sheriff deputies and campus security police escorted the FBI agents out of the crowd. The battle lines drawn, the FBI agents in their formal business suits pushed and shoved, furiously arguing with the college President, the Sheriff, and his Deputies.

The camera crews knew they had hit pay dirt and relayed the video footage as fast as they could to the BBS network. Once again, Fred Burner had the fast track on the news, and he delivered it to the nation first. He assigned a new team of technicians to help with any footage received for Colleen's hour of prime time newscast.

Colleen retreated to a studio in the Media Center and videotaped her description of exactly what had happened, describing how the FBI agents demanded that she come with them, before they identified themselves, and without giving a reason.

Colleen asked people everywhere to carry their hand-held video cameras to take pictures of any important events. She urged them to put videotapes in the hands of the BBS if they carried news not ordinarily seen on commercial TV.

More FBI agents arrived to support the demand that Colleen be placed in their custody. The Sheriff refused them access to the building and took a ranking FBI agent aside. He pointed out a large number of trucks and campers nearby, men standing among them, quietly watching.

"You see those trucks?" the Sheriff asked.

"Yes. So what?" the agent asked.

"You see the National Association of Riflemen's insignia on the trucks?"

"A bunch of red-necked deer hunters. So what?"

The Sheriff replied, "Those are local folks with children in that crowd. They don't want them hurt."

"Then tell them to go home. Tell them not to resist federal authority," the agent said.

The Sheriff gritted his teeth. "Once more, let me point out that these red-necked deer hunters are near their trucks, which they have placed all on one side of your group. They'll have a clear field of fire away from the crowd, toward where you federal heroes are standing, in case of any more trouble," the Sheriff said.

"Are you threatening me?" the agent asked.

"No, but if you had any brains, you'd see you're in danger," the Sheriff said.

The FBI agents wanted to arrest some snotty kids who humiliated them on national TV. The agents spoke of insurrection and sedition, refused to leave, and urged their superiors to send in the National Guard. The crowd around the Media Center now included students, parents, senior citizens, and feminists drawn into the fray by the BBS newscasts.

In Humboldt County, home of Arcadia State University, a 45 year-old short, wiry auto garage repairman named Earl "Early" Byrd owned several tow trucks spread around the county. The drivers constantly monitored scanner radios to learn where their services might be needed. "Early" also led the NAR rifle competition team. At sometime in the past, one of his trucks had pulled the vehicle of almost everyone in the county out of a difficult situation. When he arrived on the campus scene after the shooting, he nodded to the Sheriff who immediately walked toward "Early."

"How you doin' Sheriff?" Early asked.

"About how you'd expect with a street full of FBI warriors trying to shove their way around," the Sheriff said.

"Anything we can do to help?" Early asked. The "we" referred to a network the locals called the "Early Warning Network."

"Now that you mention it, how about asking a few of our level-headed friends to keep monitoring this situation, just like they're doing now." The Sheriff nodded toward the group of pick-up trucks nearby. "Ask the hot-heads to stay away, but near their radios," the Sheriff said.

"Awright. That's easy," Early said.

"Then ask the freeway truckers from here south to let us know if any military vehicles start moving up from the Bay Area."

"Might come to that, eh?" mused Early.

"I'll have the fire warden bring over the water cannon tanker to stand by and wash down these federales if they get too overheated," the Sheriff said. He paused in thought.

"Whatta you think? If things begin to look bad, where could we stash Dr. Jameson at the last minute? She doesn't want to go with these federal fellas, and I don't really blame her."

"Want my wife to call Auntie Susan on the Hoopa Reservation? She's got room in the Lodge. Prob'ly happy to get a visitor to talk to," Early said.

"That's a good idea," the Sheriff said. "I'll tell the Indian Marshall that the FBI might show up. Old Ben would love to block the Hoopa road with his D-9 and explain about jurisdiction to the FBI."

Early laughed. "OK. Just in case, I'll arrange to open up one of the old mine bomb-shelters in Trinity County. Everybody called people crazy to build them, so they're looking for an excuse to use them."

"That's a very good idea," the Sheriff said. "Hell of a back-up plan, Early. We should ask our taxicabs to keep full tanks so they can carry her out of here at the last minute."

"Sounds like fun," Early said. He wandered off.

52

THE BUSINESS LOBBY

As the President's limousine maneuvered toward the White House, the number of demonstrators and the amount of trash surprised him. The game plan is going well, he thought. Lots of citizens complaining while the Republicans kept a high profile in the Congress. He ought to be able to blame them for most of the dissenters' complaints.

"Sir, Senator Brown just called from California and asked you to call off the FBI," an aide said. "There is a near riot up in Humboldt County. He wants to know why the FBI is pursuing California's newest prominent citizen, Dr. Colleen Jameson. You've heard of her?"

"Sure. Berkeley virologist, won a prize. Let me check around." His aides made unproductive phone calls to the FBI, with the Director "unavailable," and Number Two and Three to the Director claiming they couldn't find him.

While he waited, the President asked his aide to call the Governor of California for a briefing. The President mused that critics called him a "goddam liberal," but called the Governor of California a "moderate Republican." How words are manipulated, he thought. His aide brought the telephone.

"How are you Governor? Could you give me a quick status report on what is going on in California. Then maybe you could tell me how this Jameson business is going up on the north coast?"

The Governor said, "I'd rather talk about that lunatic you put in at Treasury, but you asked first. Let's see. The demonstrations in the San Francisco Bay Area, Silicon Valley, and LA are large, mean, and continuing. The senior citizens are determined to get the federal laws changed. They want to recover the legislators from the lobbyists. They think that will assure their social security checks. Now they've started to complain about what they call the harassment of Colleen Jameson."

"If I pull the FBI off the Jameson case, what will Californians think?"

"Probably that you've caved in to the liberals, not that there are many in northern California. I'd rather talk about what happens if you don't call off the FBI."

"Tell me," the President said.

"I think some people of Northern California might band together to resist external authority. You recall our tax revolt years ago. Then there's the movement to split the state, north versus south. The environmental complaints have now discredited the northern timber industries. The Southern Californians contemplate running out of water, reverting to a desert, and having to learn to speak Spanish. The whole state is losing major parts of its agriculture capability. These things all contribute to citizens losing confidence in the government. If I were a Marxist, I'd say the state is ready for a revolution. Off the record, right, Mr. President?"

"Of course, Governor. Will your federal senators and representatives support the free campaign law? And how about the anti-advertising regulations?" the President asked.

"I think our legislators would like to appear consistent with our state laws. We've already passed those laws in California."

"You were saying that you don't like my Secretary of the Treasury." The President smiled. "What could be wrong? He's a Republican."

"If he defaults on all those government bonds, the Banks and the Rockefellers won't want to buy any more," the Governor said.

"In this free world market, they may prefer to invest in Chinese or Russian bonds," the President said. "Listen, I appreciate your advice. It's been very helpful, and I owe you one." They broke off.

Things sounded rather unstable in California, and the President decided to check the business climate. He arranged for a meeting with the Chair and two senior officers of the Business Lobby, all CEOs of major US corporations. He wanted to discuss the proposed federal legislation to eliminate campaign contributions and the limit on advertising.

In preparation, and on a hunch, the President had the video and audio recording apparatus dusted off, ready for the conference.

"I want a good audio record along with the video," the President said.

"You're going to record them secretly, aren't you?" his aide said, accusingly.

"Don't think about it. Nixon made it the expected thing to do," the President said, "especially among my conservative critics."

The Chair of the Business Lobby, a banker, arrived first. The President introduced Under Secretary of the Treasury Douglas, himself the Secretary until Bat Masterson received top billing.

"Bat isn't here?" the banker asked, disappointed.

Next, the head of a big TV media network arrived. "This anti-advertising attitude is childish," he began.

"I hate advertising, don't you?" the President asked, smiling and nodding as he shook hands.

The CEO of a large petroleum corporation was last to arrive. His board of directors had bought heavily into uranium ores, mistakenly thinking uranium could be used as a fuel source when oil ran out.

"Gentlemen," the President said, sitting back with folded hands and smiling. "I'd like your help in passing the anti-campaign contribution laws."

"Don't waste our time, Mr. President. We want to know why you haven't quashed this idiocy. You should be explaining to the legislators the advantages of the present Political Action Committee system."

"Just what advantages do you refer to?" the President asked.

"Surely you know that the average legislator is abysmally ignorant of the background of most issues on which he votes," the Chair said. "We provide him with knowledgeable advisers at no cost who will meet him on his schedule and bring him up to speed."

"These 'advisers' are lobbyists, paid by you to tell the legislator how to vote so as to get your contributions. You call this 'bringing him up to speed?'"

"We are just one voice among many," the banker said. "Are you trying to exclude us from any input to the policy making process?"

"Yes I am, since you drown out input from the individual citizen," the President said. "From 85% to 100% of those legislators you pay, both Democrat and Republican, vote in your favor. The public wants you out of the legislative processes."

"How do you intend to defeat the anti-advertising bill?" the media executive demanded to know.

"What's wrong with it?" the President asked.

"You don't want us to show pride in our products?" the media executive asked.

"I don't, because you own the media outlets and spend tax-free dollars to do your advertising," the President said. "Why shouldn't the taxpayer own the networks and pay directly for TV programming? Who needs the advertising?"

"You're even more anti-business than I had heard," the media executive said.

"No, I'm for the downtrodden citizen," the President said. He began to feel the businessmen's anger rising.

"We're not here to debate," the oil executive said. "If the public is unhappy with the Political Action Committees, why not scale back the amount that corporations can contribute? We can live with that."

"So you can get more of your people to make smaller payments?" the President asked. "And you still get the same amount of money into the legislators' hands?"

"We can't tolerate any kind of takeover of the media. Private enterprise has built the networks into the best information dissemination system in the world. We are the envy of the rest of the globe. Current regulations regarding advertising should be left alone. Can't you help us with the legislature? Make sure these new regulations aren't presented? Surely your prestige would carry some weight?"

"Gentlemen, I will not help you," the President said. "I suggest an alternative which might save the business community's future. Secretary Douglas, tell them what we have been thinking about."

"Yes sir. First of all, you had better give up the idea that we can get rid of Social Security payments. You are flirting with serious social dissent, probably revolt. Look out the window at the crowd if you doubt me," the Secretary said.

"Next, we strongly recommend that you send a few of your less reactionary corporate administrators to congress and try to negotiate compromise regulations. Raise the tax rate a bit on Social Security payments if you have to. But you'd better abandon the tax deductions for business advertising. You should also give up the PACs as a means by which corporations support candidates for election." The Secretary changed his tone.

"What the hell. Since you are likely to lose on this issue anyway, why not preempt the whole program. Claim that business invented the free elections and freedom from advertising plans. Submit your own bills for approval. You'll end up looking like heroes." The Secretary sat back to await the group's reaction.

The businessmen looked grim.

"I'm sorry it has come to this, Mr. President, but we seem to be at loggerheads. One last question. What do you intend to do about the broadcasts on BBS sponsored by this Jameson girl and her 'Population Institute?' Our inquiries suggest that she may be guilty of libel, slander, perhaps sedition."

"Dr. Jameson? I've enjoyed her broadcasts," the President said. "Refreshing to hear the truth for a change. You have access to the courts regarding libel and slander. I recall that sedition is resistance to lawful authority. Your Business Lobby is not America's lawful authority yet, so criticism of U.S. corporations could hardly be sedition. Your complaints begin to sound like Hitler's in the 1930s. You wish to silence the opposition, and the law be damned."

The oil executive stood up. "Well, Mr. President, Mr. Secretary, you refuse to give us any satisfaction, whatsoever. The rest of the members of the Business Lobby will be sorry to hear this."

"By the way, have you discussed these matters lately with the other members of your administration?" the Chair asked.

The media executive broke in, excited. "You are clearly out of touch, not just with the American public and the business community, but with your own administrators. I ask you again: would you reconsider the matter of investigating Colleen Jameson for misrepresentations on her TV broadcasts?"

By god, thought the President, he's threatening me with a rebellion in my own administration. Who have they turned? He thought furiously. "I'm a great believer in the First Amendment, gentlemen. You can stick your dreams of sedition up your smoke stacks," the President said.

"Well, we tried," the oil executive said, his face flushed. "You will regret today." They stamped out, grim with rage. No one talked to Big Men like this.

The President fretted. The son-of-a-bitch actually shook his fist at me. I've never seen them so arrogant. Something's going on. Is one of my staff going to confront me with a gun in his belt?

The President picked up the phone.

"You're calling the Attorney General?" Douglas asked.

"Right," the President said. When connected, he asked, "John, I've just been talking to three executives of the Business Lobby. Have you talked to them recently, about Colleen Jameson?"

"Negative," the Attorney General said.

"You didn't? Who among us do you think would most like to see her silenced?" the President asked.

Secretary Douglas muttered questioningly to the President, "Picketts?"

The President listened to the Attorney General. "Douglas agrees with you, John. I'll get back to you."

"I'll be a son-of-a-bitch," Douglas said.

"Picketts, our own FBI director," the President said. "Well, I can't say I wasn't warned. More than one person told me I'd regret appointing him."

"You think he's turned, then?" Douglas asked.

"I think these Business Lobby CEOs went around me to get the sympathy and assistance of the FBI Director. Who else, I wonder?"

When the Chair of the Business Lobby left the President and returned to his office he was angry. He called in his favorite sycophant, the Overseas Investments Director.

"Get in here, Charlie, god damn it."

"We need an early bar, is that it, sir?" Charlie opened a cabinet and threw ice and bourbon into a glass. He didn't get any. The boss didn't believe in people drinking on the job.

"An early bar and a brand new President, Charlie. I can't believe what that bastard said." The boss ground his teeth and growled as he threw down his first drink. "Who does he think he is?"

"Who, sir?" Charlie asked.

"Get out and lock the door, and get that incompetent FBI Director Picketts on the phone. Then stay outside and make sure nobody else gets on the line until I'm through. Got it?"

Charlie slid the bourbon bottle to the boss and ran from the room. The boss' intercom soon buzzed.

"Picketts?" the boss shouted. "You remember who this is?"

"Sir, I'll recognize that shout for the rest of my life. How are you?"

"I liked it better when you used to say 'Who do you want killed, now?'" the boss said.

"Who do you want killed now?" Picketts asked.

"The President. Same as last week. What's holding you?" the boss asked. "You need more money?"

"No. You sent plenty," Picketts said. "We have eight teams out now, paid in advance. One of our men will get lucky."

"Get lucky? God damn it. Picketts, get *serious* for a minute."

"Yessir."

"One day, all this will be over. Would you like our support then? I mean, you may want a job. You may need money. Do you understand me?

"Yessir."

"We need him dead. No more defaults on bonds. Oh yeah, if you get a shot at 'Balmy Bat,' take it."

"Yes sir," Picketts said.

53

ESCAPE

Back in Northern California, with Colleen still under siege, the Humboldt County Sheriff watched the FBI agents. They clustered in front of the Media Center of Arcadia University. Perhaps two hundred students, assorted faculty, parents, and a few deputies barred them from access to Colleen in the building.

The Sheriff spoke to his deputy nearby in a voice loud enough for the FBI agents to hear. "Have you ever seen FBI agents stand around and just wait like this before?" He wondered if he could provoke the agents into leaving.

"Nope," the deputy said. "They look like a bunch of waiters with those black suits. You reckon they don't have homes to go to? Don't they know how bad they look on TV?"

"Yeah, and they don't even have any nightsticks to beat students with," the Sheriff said.

With this, a nearby FBI agent detached himself from his group and walked belligerently toward the Sheriff. "I am Agent Brownlee, Sheriff. Do you find this situation amusing?"

"No," the Sheriff said. "I find it disgusting. Why don't you just give it up and go on home? Where are you stationed, anyway? I don't recognize you."

"We are under orders to detain Jameson for questioning. She is not to be allowed to make more newscasts. If you don't help us, Sheriff, you'll be held responsible."

"Dr. Jameson lives in this community and is a law abiding citizen. Has she irritated the Washington political establishment, Agent Brownlee?"

"We have orders to prevent her from continuing her newscasts. We intend to take her into custody."

"Isn't that a violation of Dr. Jameson's right of free speech, Agent Brownlee?" the Sheriff said, sarcastically. "It seems we don't have anything to arrest her for, Agent Brownlee."

"We will arrest her, Sheriff," the agent said, finally enraged.

The Sheriff's demeanor changed, and he spoke in a low, serious tone. "Who gave you these orders, son? Can't you go make a telephone call and check with the next man up the ladder? You are making a serious mistake, and I think you know it. C'mon, I'll let you use my car phone." He placed his hand lightly under the agent's arm and pulled him gently toward the patrol car.

"Get your hands off me, God damn you. I'm under orders!" The agent jerked his arm loose and stormed back to his group.

Oh my, oh my, the poor bastard's wound up tight, the Sheriff thought. Somebody wants Dr. Jameson, and they really put a fire under those boys. If they thought they could make it, they'd try to storm through the crowd and drag Colleen off in cuffs, screaming to the mob. Boy, I'm glad I'm not them.

He turned to his deputy. "These boys are acting pretty strange. Go find 'Early' and tell him that we may have to move Dr. Jameson pretty quick."

Shortly after Humboldt County Sheriff Scotty Weiss sparred with the FBI agents to learn how badly they wanted Colleen, Evan Thornton arrived on the scene and worked his way through the crowd into the Media Center. He finally got into the interview studio and found Colleen. He put his arms around her and kissed her.

"My God, I worried about you," Evan said. "Are you all right?"

Colleen disengaged herself from him and said, "I don't want all this public display, Evan."

Evan shrugged. "If you were mine, I wouldn't let you act so foolishly. Why are you holed up in here? Why not talk to the FBI outside?"

"I'm not yours, Evan. And I'm in here because I don't trust them. Why do they want to talk to me?" she asked.

At this moment a small, plainly dressed woman with a shawl over her head came up to Colleen.

Evan looked at her strangely and pushed her away. Colleen ran to the woman to reassure her.

"Could I talk to you?" the woman asked as she drew Colleen to one side. Evan tried to follow.

Colleen now felt afraid of Evan. "Are you trying to protect me?" Colleen asked. "Go find something to do, Evan." She turned to the woman tugging at her.

"My name is Tina Lauder, and I live on the Hoopa Indian Reservation a few miles from here. The Sheriff just talked to me and asked me to come and find you. He thinks something is wrong out there with the FBI agents. He is scared they'll try to force their way into the building and take you away."

"Do you trust the Sheriff? What does he say we should do?" Colleen asked.

"Oh yes. He's one of the few locals the Hoopa tribe does trust. He made arrangements to sneak you out of here if you want. I can take you to our lodge on the Reservation. The FBI or nobody can come onto the Reservation. If you don't want to go there, the Sheriff made arrangements to take you away to a bomb shelter."

"A bomb shelter?" Colleen asked, startled.

"Yeah, well its really nice, sort of an underground house. That's where I'd go if I was worried. Anyway, you can decide later. I'll wait in case you want to go with me."

Colleen impulsively gave her a kiss and went back into the recording studio. She thought, If I'm going to be abducted and disappear, I'm going to warn people ahead of time. She recorded a short videotape to leave behind.

She described the latest events, and said she might have to place herself under the protection of the local law officials to save herself from the FBI agents. The agents refused to leave the campus when asked by the College President and the local law men. She urged the listeners to keep fighting for the new legislation and signed off.

"What did you mean just then?" demanded Evan, who listened to her broadcast. "You trust the local law more than the FBI?"

"You're damned right, Evan. That's just what I mean."

Evan grabbed her forcefully by both arms. "Colleen, don't do this. You'll end up a fugitive, a wanted criminal."

Colleen stiffened. "Take your hands off me, Evan."

He dropped his arms to his sides.

"Don't let yesterday give you any ideas, Evan. You don't own me." She glared at him.

Tina once more tugged at her arm. She whispered in her ear, "C'mon to the bathroom." Inside, she said nervously, "Sheriff says some trucks with soldiers are coming from San Francisco. It's time for you to go." She removed her shawl and unfolded a long skirt hidden under her blouse. "Put these on." She held out a safety pin.

Colleen finally realized the seriousness of the situation by seeing this woman's concern to get her out. The skirt and shawl hid her clothes and face. Tina looked her over. "Nobody will recognize you. Go on out the front door and walk slow over to the taxi waiting on the street behind the fire truck. Keep your head down, maybe limp a little. You better go now," Tina muttered urgently.

"My god, Tina. Thank you." She hugged her, then walked slowly to the door and into the crowd.

Blinded by the lights as she stepped out the door, Colleen remembered to limp. The FBI agents looked at her. She saw a taxicab's light in the street and barely restrained herself from running.

"Stop right there," shouted one of the agents.

Colleen froze, and heard running feet. An outbreak of angry accusations broke out behind her in the crowd, and she started walking again. As she reached the taxi, and opened the door, the driver said, "Let's go."

Colleen jumped in and pulled the door to close it. To her dismay, Evan pulled the door open and jumped in with her.

"I'm going along to take care of you," he said.

Colleen couldn't risk an argument. "Let's go," she said. The driver slowly accelerated, and the taxi retreated from the crowd and the campus into the darkness.

Colleen shuddered with relief.

54

PLOT?

Soon after Colleen's taxicab left the campus, the advance scout car for a military convoy pulled up near Sheriff Weiss' truck. A major with sidearm hopped out of the car, ignored the FBI agents, and walked briskly to the Sheriff.

"I'm glad they only sent a major, " the Sheriff told a deputy. "A general could escalate this into a full scale war."

"Greetings, Sheriff," the major said, looking around. "You must have tight control here to have prevented any further bloodshed."

"We try," the Sheriff said. "This is a tight-knit community, with a lot of volunteers, some you can't see outside the lights. We have a lot of fishermen and hunters who live around here." He nodded toward the campers and fire and tow trucks, hoping the major would understand what he meant.

"We're concerned to keep the crowd of civilians from being

hurt. They're still too mad to go home, so I want to give them time to cool off. Now, what can I do for you?"

"Maybe nothing," the major said. "My orders are to help keep the peace by making myself available to the local authorities. That's you, I think."

"I appreciate your assistance," Sheriff Weiss said. "How many troops do you have?"

"About thirty, a mixed bag of those available when the call came through. They have rifles or sidearms and four vehicles. They are equal to standing guard duty, and not much more."

"Then you're not the regular National Guard?" the Sheriff asked.

"No," the major said. "My commanding officer responded to a call from Washington and told me to gather available personnel and get here quickly."

"Major, if we are going to act in coordinated fashion, I have to tell you what I am trying to do. You'll have to decide whether you trust me or not." Let's see if he believes that, the Sheriff thought.

"That's fine, Sheriff. Let's hear it," the major said.

"Before the shooting, FBI agents approached Dr. Colleen Jameson and tried to arrest her. However, this community and the students like her. The students have a saying: Religion is the opiate of the masses and Dr. Jameson is the heroine of the students."

The major burst into laughter. "This promises to be interesting. What happened then?"

The Sheriff speculated. This guy caught my reference, so he didn't come up through the military ranks. Also, he's not here officially. Who in Washington sent him? Sheriff Weiss determined to find out.

"Dr. Jameson retreated into this building which houses a lot of video equipment. The operators turned on the cameras, caught the mob scene and the shooting, and made the FBI look bad. Now the agents are mad as hell. How about asking them why they don't leave? We can compare notes."

The major walked into the center of a group of agitated and gesticulating FBI agents, and was told that they would take "the Jameson woman" into custody, and, damned if they'd call her "Dr." They talked to the Major earnestly.

The major returned. "I agree. They're psyched up and want her bad. But isn't it possible they're just trying to do their jobs?"

"Let's hope so. I may be worried needlessly," the Sheriff said. So the major wants to reassure me? There isn't any problem at all. How nice, Scotty thought.

The army vehicles parked and disgorged an assortment of troops. Three women noncoms looked as competent as the rest. A finance corps officer nervously fingered the handgun on his hip. What was that thing anyway?

The major placed a group of soldiers on each side of the FBI agents and told them to look commanding. They needed full battle gear with rifles. Somehow, all this disarray reassured the Sheriff. He called Early to orchestrate the show.

"Early, could you talk to the guys who have laser sights on their weapons? Once in a while, have them shine the light beam across the faces of the FBI and the major. Let 'em know they are often in our thoughts and sights."

"Sheriff, you think like the commanding officer of my old rifle company. You ever spend time in combat?" Early looked at him, amused.

"I'll buy you a beer when this is over and we'll compare notes," the Sheriff said. "I'll tell you one thing though. My field officers were good but nowhere near as good as you are."

Early grinned. "Thanks, Sheriff." He walked quietly into the shadows.

As the evening wore on, the FBI agents got tired and more obstreperous. Their senior officer came to argue with the Sheriff.

"Move this crowd out, Sheriff. We want to search the building."

"These citizens have a right to assemble peacefully. I can't force them to move," the Sheriff said.

"We'll find her eventually."

Suddenly a shout went up inside the building. An FBI agent had disguised himself by removing his coat and tie, and mussing up his hair. He worked himself into the crowd, then into the building.

"She's not here." The agent stormed out of the building. "You lied to us, Sheriff. You're concealing her. Where is she?"

The Sheriff simply shook his head. Then he heard the words, "She's on the damned Indian Reservation."

Who told them? the Sheriff wondered. He found Tina, but she denied telling anyone but Colleen about the Reservation. Then who did tell? he wondered. Well, I'd better get there first, he thought. He jumped into his truck and took off. He telephoned the Indian Marshall as he drove.

"Ben, I'm sorry, but a bunch of crazy FBI agents and a few dozen soldiers are looking for our newest citizen here, Dr. Colleen Jameson. I think she is headed for Red Valley, but they think she's in Hoopa. They'll probably threaten to shoot you if you try to stop them. That poor lady has driven everyone crazy, telling the truth about who's really running this country," the Sheriff said.

"Boy, that'll get her in trouble," Ben said. "What do you think I should do?"

"Tell people I said you should close the road. These guys worry me. Use the big caterpillar. Hold 'em off until they find out she's not on the Reservation. But you'd better do it now. They'll be at your place in about 45 minutes. I'll be ahead of them. Listen, Ben, I'm sorry about this."

"Don't worry about it, Sheriff," Ben said. "We've been through this stuff before, haven't we?" Yes, he has, the Sheriff thought, regretfully.

Early called him with the latest news. "The FBI is loading two cars with agents and the major mounted up his troops. They'll all take off for Hoopa or maybe Red Valley any minute now."

"Dr. Jameson went to Red Valley, I guess?" the Sheriff asked.

"Yeah. I talked to my cousin, the Civil Defense Coordinator. He'll put her up in a small survival shelter, an old mine. Then they'll mobilize their network if need be. Boy, those Red Valley folks really don't like the feds pushing us around."

"That's good. Thanks, Early," the Sheriff said. "By the way. How did the feds find out that we might hide her in Hoopa?"

Early rubbed the stubble on his chin. "I can't imagine any of us would tell them. Has she got friends who aren't on her side?"

"Maybe."

While the FBI pursued Colleen in California, the President barricaded himself in the White House. He resented the people throwing trash around the grounds, but he agreed with their cause.

He called Douglas of Treasury and his National Security Adviser and asked them to bring a couple of aides. When they arrived, he described the problem.

"First, I'll tell you several things I believe to be true. Then I want you to give me a scenario consistent with the information I give you. The scenario will describe how certain people would try to remove me from the Presidency, perhaps by assassination. OK?"

The National Security Advisor promptly said, "Damn it." A sharp intake of breath came from his aide. "Let's hear it. Damn," he said, again.

"OK," the President said. "Last week, I told the executive committee of the Business Lobby that I would fight them as long as they try to save PACs and tax subsidized commercial advertising.

They told me, 'You will regret this. Members of your own administration disagree with you.'"

"I asked the Attorney General if he had talked to these business bastards. He had not, but thought Mr. Picketts, head of the FBI, a pretty good candidate. Picketts has said in public that I am too soft on dissenters, among other things. Right, Douglas?"

"Yes sir. I heard Picketts froth, myself," growled Secretary Douglas.

"Next data. The national commercial television channels all denied access to Secretary Douglas' precocious understudy, our own Bat Masterson. Bat had to approach Dr. Colleen Jameson who has an arrangement with Fred Burner for a TV newscast. I infer that the Business Lobby told its members to do all they can to isolate and destroy this administration."

"Third disquieting fact. The Sheriff of the California county where Dr. Jameson runs her Population Institute told the Governor that a group of FBI agents is trying to take her into custody for sedition. Only the students saved her."

"Those bastards," said an aide, a young woman who followed Colleen's progress avidly. "Whose authority is she resisting? Picketts? Sedition, my ass."

"Exactly," the President said. "Next, the Sheriff reported that three dozen or so assorted soldiers showed up to help him 'keep order.' The ranking military officer said his commanding officer sent him up from the San Francisco area, at the request of someone in Washington, D.C."

"At this very minute the military contingent and the band of angry FBI agents are looking for Dr. Jameson. What they will do with her if they find her, God only knows. The only bright spot in all this is the Humboldt County Sheriff and his associates. I'm told he'll make sure they do not find Dr. Jameson. I'll talk to him later."

"My first question is, who gave the FBI agents their marching orders, and what are those orders? To detain Dr. Jameson, as they say? The second question is, who sent this rag-tag military band, trying to make it look like the National Guard? And what are their orders? Now you guys talk for a while." The President leaned back and swiveled in his chair, waiting.

"It seems likely that Picketts ordered the FBI agents to pick up Dr. Jameson," the Security Adviser said, "on a fabricated but serious charge, so they could keep her off the air for a long time. Is forever a long time?"

"Yes, and I agree," Secretary Douglas said.

"Next, who is the most likely person in the military who would want to send out a team?" the Security Adviser asked. "And for what purpose?"

"They can't very well take her into custody. Would somebody in the Business Lobby like her dead?" an aide asked.

"To shut up the BBS newscast? Absolutely," the Security Adviser said. "Otherwise, they may lose every penny they've stolen through their past business practices."

"Who would the Business Lobby's CEOs go to, to assemble a military assassination squad?" the President asked.

"To their old friends in the Pentagon. They've been collaborating on defense contracts for decades. Some of those generals and admirals hope to be CEOs in the business community, if they're ever driven out of the military," Secretary Douglas said.

"All right," the President said. "We have motive to remove me. We have malcontents close to me in Washington, such as Picketts. Where would they approach, to force me out?"

The National Security Adviser said, "Right here in the White House. They know where you are when you're in town. And there is the cover confusion of the demonstrations outside."

"You all agree? This is a plausible scenario?" the President asked.

Slowly, they all nodded.

"I think so too." the President said. He sat and let this sink in. "So I'll travel for a while. Tell people I'm barricaded in the White House." He got up in a thoughtful mood, and started to leave.

The young aides sat devastated. "Damn it sir. Isn't there something we can do?" one asked.

"I can't think of anything, when dissent this serious has reached so high among us. If you do, let me know," He looked at the stunned aides and sat back down.

"You mustn't worry," he told them. "Right now we have a chance to make some very important changes. We really might save the world. Or, more likely, Colleen Jameson and her colleagues may save the world." He leaned intently toward the young aides. "How could we have been so lucky as to have participated in such momentous events?" Then the President smiled at them and sent them away.

"Will you hold the fort while I'm gone?" the President asked Treasury Secretary Douglas.

"Do my best," he said.

"I'm going to take some time to get to know Colleen Jameson," the President said. He called his Attorney General, John MacLalen.

"John, I have something that I think is important. I'd like you to take a trip with me for three or four days. We leave in three hours."

"Can't possibly, sir. I have a big organized crime conspiracy case."

"I'll call you back in a minute and give you the name of another person to bring. It's important."

MacLalen answered immediately. "All right, sir. I'll be ready."

Then the President called his newest appointee to the Supreme Court.

"Justice Suggs, forgive me for bothering you, but I have an unusual request."

"How are you Mr. President? How can I help?"

"I would like to borrow one of your law clerks for a few days," the President said. "If things go well, we will have wasted your clerk's time and he will tell you about going fishing for steelhead."

"Mr. President, this seems somewhere between unwise and improper," Justice Suggs said. "We certainly won't want any future accusations that you tried to exercise undue influence on the judiciary."

"We most certainly do not, your honor. I'm more interested in an unimpeachably objective observer who can give you first-hand information if it should become necessary."

"You mentioned what would happen if things go well. What if things do not go well?" the Judge asked.

"I fear we may watch an illegal takeover of the government," the President said.

The judge was silent for a moment. "Well, we certainly don't want that. Who should my clerk call for arrangements?"

"Attorney General MacLalen, sir. And thank you, Justice Suggs."

"You're welcome, and good luck."

The President awoke to shots fired at some distance. He listened to the shouting outside his room and got up to look out his door.

His aide appeared. "What's up?" he asked.

"Someone outside fired shots into one of the rooms. I'll be back." He turned on his heel and ran.

The President watched a guard with a rifle peering outside into the illuminated yard. When a door burst open, the guard whirled raising his rifle in readiness. For that moment the President felt the emotions of someone standing empty handed while another confronted him with a weapon.

His aide returned. "Someone showed up outside the fence with an automatic rifle. He got off a full clip, chucked the rifle, and jumped in a waiting cab. The police are chasing cabs now."

As the President settled himself in his bed he found his thoughts returning to the period he spent in infantry basic training, a rifle with him most of the time. Maybe I should reconsider my position on gun control, he thought.

55

THEY ALL SAY THEY LIKE IT WHEN IT'S OVER

Colleen's taxi quietly stole her away from the Arcadia State University campus under cover of darkness. The occupants found themselves on a winding, narrow, two-lane paved road that took them over the coastal range of mountains and back down into the interior valley. For two hours they didn't see another car. The lights swept from one side of the road to the other continuously as the road twisted its way through the timbered hills.

The bulky driver wore a cap and a leather jacket to keep out the weather. She wore no seat belt, didn't seem to care about the jostling back and forth, and used the brake often. Matter-of-fact and cheerful, she laughed easily. Colleen wondered how she could be so placid, fleeing from law enforcement officials.

Colleen leaned forward to talk to her. "I really appreciate this middle-of-the-night help."

"Oh that's OK. You're Colleen? What's going on, anyway?"

"I really don't know. It may be that the TV program has created some enemies."

"Oh, yeah," the driver said. "I like your programs. They're so good they probably make some people mad. Are they fun to do?"

"Well, it's fun to tell the truth in public," Colleen said. "There's a lot of work getting the programs ready."

"Got any kids?" the driver asked.

"No. How about you?" Colleen always winced when a woman asked her this question, but she'd be damned if she would say "Not yet." Our society really pushes women, she thought.

"One," the driver said. "Enough to make me need this job. I see my husband at breakfast and dinner."

"What will happen when the troops go looking for me in Hoopa?" Colleen asked.

"Nothing," the driver laughed. "They'll never get to Hoopa. The Indian Marshall will stop them."

"Is the Indian Marshall an Indian who is a Marshall, or is he a Marshall for the Indians?" Colleen asked.

"In this case, both. The Hoopas won the right to have their own Marshall for the Reservation. Ben is a full blooded Native American Hoopa," the driver said. "We had lots of jurisdiction battles before. The reservation follows along the river's path, and the Indians used to put up gill nets and cables to block the river. The sports fishermen stayed mad at them. Blamed the Indians for taking all the fish even though the offshore commercial boats took most of the salmon. It's peaceful now. Not happy, but peaceful."

A few lights appeared in the distance. "I'm going to stay out of town and go directly to the shelter," the driver said.

"How will the Hoopa Marshall stop the troops from going into the Reservation?" Colleen asked.

"The road is narrow, and he can block it easy. When they finally decide that you're not there, they may follow us here."

"Would the local police tell them where we are?" Colleen asked.

"I don't think so. These FBI guys have made a very bad impression, and there are a lot of survivalists and militia members where we're going," the driver said.

"Survivalists?" Colleen asked.

"You know, people who worried about nuclear war and other disasters. California taught civil defense courses for a long time, how to build bomb shelters, and all that junk. Lots of people up here took it very seriously and stored food away."

"You also mentioned militia?" Colleen asked.

"They're the people who really don't trust the federal government. They believe in self-protection, and nobody has a right to govern them above the Sheriff at the county level, that sort of thing. You're going to a bomb shelter that is very well equipped, very comfortable. It's in an old mine."

Evan started to retch, carsick. The driver pulled to one side and stopped. Evan staggered out and fell down on his knees. His wallet fell on the ground. Colleen found herself annoyed with his loss of control. She grabbed the wallet and shoved it in her purse.

As they approached the town, the driver slowed down. She inquired on the radio for directions, and headed for the shelter, a few minutes outside of town. When they left the paved highway, they traveled down a narrow dirt road until they reached shacks left over from the original mine operation. The taxi's headlights showed a pick-up truck and a woman who waved at them. Colleen's driver stopped. They all got out. The driver pulled Colleen to one side and whispered to her.

"I don't know how well you know this fella you're with, but I think he deals drugs."

Colleen froze in horror. "Oh god, no. Are you sure?"

"Pretty sure," the driver said. "I think he had me take him on a delivery in Arcadia night before last. Listen, good luck." She patted Colleen's arm and got into her taxi, smiling. "I'll get home just in time for breakfast with my husband."

The woman who waved at them got out of a truck. She wore a deputy's badge on her green jacket.

"I'm Doris Price, and I'll get you settled. Brrr," she said. "Let's go on in. It's cold out here."

They approached the entrance of a tunnel, closed off by two heavy steel doors welded with reinforcement bar hinges, set in concrete on each side. The deputy opened one of the doors, and Colleen had her first touch of fear, or claustrophobia.

"Some people don't like mines. But I'm used to this one," the policewoman said. She flipped on a light, and Colleen found herself looking at a mobile home fitted into the tunnel. Deputy Price opened the door, and they stepped up into the living room. From the inside, one couldn't tell they were underground. "Are you tired?" The deputy looked at Colleen and ignored Evan.

"Exhausted," Colleen said.

"Well, there's two bedrooms back there. Chief Denning asked me to stay here until you get settled. I'm on graveyard shift, so I'm not tired. I'll stay out here for a while, and you can get some sleep."

Evan tried to join Colleen in her bedroom. She resisted, still afraid of him.

Colleen pleaded, "Not here," and shushed and pushed him out of the bedroom. She frantically fastened the latch on the door thinking about what the taxi driver told her. A drug dealer? As Colleen readied herself for bed, she remembered Evan's wallet. She looked at the inadequate latch on the door, then pulled his wallet from her purse.

She riffled through a thick sheaf of bills, mostly \$50s and \$100s, over \$4,000. She assumed it was drug money, and her stomach contracted in fear. Next came two credit cards with the name Evan Y. Rose. So he gave us a false name. What's next? A condom surprised her. She assumed he didn't know what they were. California driver's license, issued two months ago; FCC Radio Telephone Operator's permit; and a permit to carry a firearm issued in Washington, D.C. What is he? she wondered. A policeman? Then she saw two thin packets of white powder between two credit cards. She thought she saw the door move inward as the latch made a small noise.

Her hands shook as she jammed everything back in the wallet and shoved it in a drawer. She turned off the light and crawled in bed. Anxious, but exhausted, she went to sleep.

Something woke Colleen. She listened and heard a vehicle pull away, accelerating as the driver shifted up through the gears. Then she began to wonder who left. The policewoman, Doris Price? A stab of fear went through her. She lay in the dark hoping she would go back to sleep and somehow wake up in the morning with everything all right. Every creak in the trailer increased her dread until she decided to go out to the kitchen and make some coffee, to wait for morning. Where was the can of mace she had always planned to buy for her purse? Still in the store.

She put on slacks and a sweater, screwed up her courage, and unlatched the door. She swung it open to find Evan standing a few inches away, in his pants, without a shirt, a hand on each door jamb, flexing his muscles, leering at her.

Colleen screamed in surprise and fright and tried to close the door. Evan easily pushed through and sat down on her bed. The small room scarcely gave her space to stand at the foot of the bed.

"Come on. Sit down." He made room for her, patting the bed.

She recoiled. "Get out, Evan." She narrowly avoided accusing him of using a false name with her. "I don't want you in here."

"Aw, come on, Colleen. You've kept me away from you all this time. I miss you." He reached out to pull her to him.

Colleen stepped sideways out the door, avoiding his hands. She ran to the kitchen, saw a wooden knife block, and pulled out two paring knives. She grasped each handle so the blade pointed backward toward her elbow. With her thumb she held them concealed in her palm. She defiantly turned to face Evan, not sure whether he saw her take the knives.

"I don't want you to touch me, Evan. If you want to talk, then sit down, but don't come near me." Colleen discreetly looked at her watch. Six a.m., and the sun should come up soon, but she had no way of knowing when the Police Deputy would return.

As Evan reluctantly sat down, Colleen searched his face. She looked for something to help her accept that he was a drug dealer with a false name who legally carried a gun in the nation's capital. Nothing fit. His words said he cared for her, but his expression seemed deceptive and contemptuous. She decided he could hurt her.

"Please don't push me away, Colleen." He spread his hands. "I'm falling in love with you."

Colleen tried to keep him talking, anything to keep him from getting physical. "Funny way of showing it, Evan. You tried to control me at the University last night, and you pushed your way into the cab."

"Do you blame me for trying to stay close to you? I want to protect you. Frankly, I think you'd have been safer going with the FBI in Arcadia than out here in this hole in the ground. A hidden bomb shelter, for God's sake, in the hands of a hick town police force," Evan said.

"What do you think the FBI wanted with me?"

Evan shrugged quickly. "How would I know." He looked away. "Probably wanted to ask you some questions."

"About what?" Colleen asked.

"Well, maybe your broadcasts," Evan said.

"Why would the FBI care about what I say on TV?"

Evan looked incredulous. "You don't see anything wrong with destroying people's confidence in American business?"

"If you didn't like the program content, why did you come to work for the Institute?" Colleen asked, trying to draw him out.

"It sounded exciting at first. Then I got interested in you. You're really important to me, Colleen." He started to get up.

"Leave me alone," warned Colleen, standing up.

Evan smiled playfully. "What'll happen if I don't?" He stood up and walked close to her, leaning down as if to kiss her. "What will happen?" He slowly embraced her, and she put her arms around him. "That's better." He started to kiss her, saying "See, nothing's going to happen."

"What are you doing?" she asked, turning her head to one side.

"I want to make love to you. I really care about you, Colleen."

"But I told you not to come near me and not to touch me like you're doing now," she said.

"It feels so good to hold you. I can't let you go." He buried his face in her hair and kissed her neck.

"Evan, let me go. Now." Evan ignored the menace in her tone. Maybe he likes it, Colleen worried.

"I'm going to take you in the bedroom and enjoy you." He whispered in her ear, "I'm going to fuck you."

"You're going to rape me?" Colleen asked.

He whispered again, "I don't care. I have to have you. You'll like it. Wait and see. When it's over, they all . . ."

"They all like it, do they, Evan? So you're going to rape me?" Colleen asked. She quickly slanted the paring knives away from her wrists where the blades had been concealed and pressed each point against Evan's back over his kidneys. The blades pointed inward to bury in his back if he tried to push her away.

"Ahh," Evan screamed. He struggled for a moment, which forced the knife points into his skin and increased his pain. "Please, Colleen, stop," he screamed.

"You said you wanted to rape me, Evan. And I would enjoy it even though I asked you not to."

"Please, Colleen. You're hurting me."

"No, you are pulling the blades into you, trying to get away. Don't move, Evan."

He tried to hold still, but shook convulsively. "Please Colleen."

"Clasp your hands behind your head." He quickly did so.

"Now, be very careful and don't move, Evan. I'm going to take away one hand, but the other will keep one knife in your back. If you move, that knife goes in. Do you understand?"

He nodded and whimpered now, his face sweating.

She quickly moved her right hand from behind his back and placed it down between his thighs, with the knife blade pressing up into his scrotum. He stood frozen with terror and screamed full voice.

"Evan, tell me you'll never rape another woman. Otherwise, I'll fix you so you can't."

"Never, Colleen. God in heaven. Never again. I promise."

"And you won't ever touch me again, Evan? Never again, right?"

"Never, Colleen. Honest. I'll never touch you again," he blurted out the words he hoped she wanted to hear.

Colleen realized he had messed his pants. She also heard a car approaching from the road.

"All right, Evan. When I take the blades away, go in the bathroom and clean yourself up. Never touch me again, Evan?"

"Right," he said, totally immobile. "Never again."

Colleen slowly took the knife blade away from his back He tried to stop his wailing as he backed gingerly away from the blade of the paring knife poised under his genitals. Once free, he turned, weeping, and lumbered into the back of the trailer to the bathroom, two small wounds bleeding on his back.

Colleen marveled at how well the kitchen utensils pacified Evan. The small knife served her so well she tucked it in her purse.

Deputy Doris Price parked her pick-up and carried a bag of food into the kitchen. Colleen hugged her with relief. Doris looked toward the bedrooms.

"Anything you want to tell me?"

Colleen started to speak, then shook her head, "No."

Doris started cooking coffee and bacon and eggs. Her presence and the cooking smells and sounds transformed the kitchen into a comforting scene. Colleen heard the shower running a long time before Evan returned to his bedroom.

Colleen took her coffee outside and shivered in the cold air as she watched the sun come up. The rays of light slowly illuminated the surrounding clouds and hills and trees, and brought her back to the comfort of the everyday world. I've got to get back to the outdoors she promised herself as the night's terrors faded.

56

Back at Arcadia State University, the FBI learned that Colleen had evaded them. It took time to load their cars and find directions for the pursuit. The two FBI cars reached the Hoopa road before the military trucks. The road became progressively narrower, the trees on each side grew closer together, the branches extending into the road. Ahead, the agents saw a large tractor blocking the road.

The first car pulled up and an agent jumped out, gun drawn.

Sheriff Weiss waited behind the tractor with the Indian Marshall. The Sheriff left his truch earlier back on the main road and walked in to Ben's tractor.

"This is Sheriff Weiss and Indian Marshall Ben Shoals," the Sheriff shouted. "You have no jurisdiction here. This is an Indian Reservation. Dr. Colleen Jameson is not here. She is back in Arcadia."

The FBI agent let out a cry of frustration and fired a shot at the tractor blade. The shot infuriated Ben, and he shouted:

"This is Hoopa Indian Land. Get out of here. You have no . . ."

His sentence stopped as the FBI agent fired another shot. The bullet hit Ben, and he sat back down on the tractor.

"Ben, where are you hit?"

Ben answered clearly, "Bullet scraped my ribs, I hope."

"God damn you crazy fools," shouted Sheriff Weiss. "I told you Colleen Jameson is not here. You had no reason to shoot this man."

The agent laughed. "We're in hot pursuit of an accused felon, and you blocked our way without identifying yourselves."

Another agent called out. "Come on, let's get out of here. She's over on the other side of the mountain."

"Unblock this road," called the Sheriff. You clear it or we will." The Sheriff jumped into the tractor seat, started the engine, and headed the mechanical giant toward the FBI cars. He raised the blade as a shield. The FBI drivers frantically backed up, weaving and cursing, until they got to the main road. They headed toward Red Valley.

A truck followed the tractor. It stopped and Ben's friends got out. "We'll take Ben to the hospital," one said.

"All right. I'll follow these bastards until we can stop them."

The Sheriff jumped into his truck and immediately called his dispatcher. She broke in, excitedly.

"Sir, do you remember yesterday when you talked to that person on the East Coast? Very highly placed person?"

"Yes, of course I do." Sheriff Weiss had talked to somebody who said he was on the President's staff.

"That person called and asked you to meet him and his boss at the Red Valley Airport tomorrow morning. They want to meet Dr. Jameson."

"His boss? That's what he said?" Sheriff Weiss hoped this didn't mean what he thought it did.

"Yessir."

"OK, I'll be there. Now, would you wake up Police Chief Denning in Red Valley and ask him to call me. Tell him what's happened and where I am. Then have Early call me. OK?"

"Yessir."

As they talked, the Army trucks careened past his vehicle, lights blazing. The Jimmy turbochargers whistled over the roar of the two-cycle diesels. The Sheriff joined the caravan with its weary FBI agents in front, and peace-time soldiers jumbled about in the swaying trucks that brought up the rear.

When Police Chief Denning called from Red Valley he had been asleep for several hours. "What's up Scotty. I saw the shooting on the BBS."

"Yeah, Chief. I'm ashamed that I didn't see it coming. They say the girl will live, at least. I wanted to tell you something that bothers me about this whole operation."

Sheriff Weiss described what had just happened at Hoopa, Ben's shooting, and the string of vehicles with their bizarre occupants that he now followed.

"Chief, have you ever seen FBI agents act so nervous and uncertain? These guys seem very excited and jumpy. Belligerent. The military group is just a bunch of peacetime soldiers who happened to be handy when they got the call. They aren't professional soldiers with combat training. All except for the ranking officer, a major, who looks like he's fresh out of Fort Benning paratrooper school. Very cool, and a smartass. Tried to tell me not to worry."

"This is the bunch coming over now?" Denning asked.

"Right. Now you could decide to shake out all your people and try to get a bunch of State Police in there to meet these cuckoos when they arrive. You could demand their weapons, but I think you might end up with a shooting match. The FBI would eventually open up with their oversize magazine .41 autos. There's nine or ten of them, so they could put out a lot of firepower. They seem paranoid to me. They are bound and determined to find Dr. Jameson."

"What about the soldiers?" the Police Chief asked.

"Fortunately, they're getting more and more carsick as we go along, throwing up over the tailgate."

"So what did you have in mind?" the Chief asked.

"Could you meet these guys and invite them somewhere, maybe to the old armory?" the Sheriff asked. "Offer them some coffee and a sandwich. Let them use the restrooms. Tell them to rest while you put out search crews to find Dr. Jameson. Of course, you'll tell them the minute she's located."

"Well, that's not hard to arrange. Do you think they'll accept the offer?" the Chief asked.

"You have three women deputies don't you? Get them to meet this convoy and escort everybody to the armory. Feed 'em and talk to 'em. Might stand a better chance of pacifying these folks. The FBI agents are full of determination."

"That sounds very sensible, Scotty," the Chief said. "Then we play it by ear? Try to split them up, and get their guns away from them?"

"Something like that. Oh yeah, one more problem. I got a call from somebody in Washington, DC, yesterday, the White House, if you can believe it. One of the President's aides wanted to know the FBI's mission in Arcadia." The Sheriff waited for that to sink in.

Chief Denning asked, "Well, why don't they just call the FBI Director and ask him?"

The Sheriff said, "Exactly. I thought about that for a while and got scared. That's also part of why I'm calling you. Something very odd about this whole situation."

"Scotty, I don't like what you're telling me." The Chief's voice now showed alarm. "What you're saying is that the FBI and the White House might not be the best of friends at the moment?"

"I'm glad you said it," the Sheriff said. "I thought I might be crazy. Now all of this has to be between you and me, Chief, OK?"

"Jesus Christ," the Chief said, muttering his dismay.

"I got a call a few minutes ago asking that I meet a plane at the Red Valley Airport in the morning. Somebody is coming in from the White House, and they asked me to meet them. They want to see Dr. Jameson. I think you ought to come with me, without any of your people knowing about it."

"I wish I had recorded this so we both don't fry later," Chief Denning said. "OK, you got me curious. I'll be there. Anyway, we have to figure out what to do with these people we want to get into the armory. Do we arrest them?"

"If you could. One last thing," the Sheriff said. "I asked Early Byrd to alert his friends on our side and your side of the mountains to arrange a hiding place for Dr. Jameson."

"Yeah," the Chief said. "He called to tell me, and it's all set."

"Good. I'll see you in the morning at the airport."

Sometime after Sheriff Weiss finished talking to Chief Denning, the convoy slowed to a stop ahead of him. The soldiers leaped out of the truck and ran into the brush along the road. When finished, they crawled back over the tailgate, and the truck moved out. As Weiss's truck moved forward, a woman jumped out into his headlights and ran back to the door on his right side. He saw her second lieutenant's bars glisten.

"Please let me come with you," she said urgently.

"Hop in." What next? wondered the Sheriff.

She sat shaking from cold and fatigue. "Thank you. The rest of us in the truck voted to send someone to tell you that we aren't prepared to use our weapons."

"That's good. It could save lives," Weiss said.

Words poured out of the woman. "As the day wore on, the whole operation looked more and more unplanned. Our finance lieutenant asked the major why he didn't get a bunch of military police or somebody, instead of us. The major said he didn't have time. We don't want to disobey orders. We may pretend to follow orders, but we certainly aren't going to shoot anybody."

"Tell me what you know about the major," Weiss said.

"Wally? Oh, Major Deevers. He's new to our headquarters. He took me out for drinks a couple of times. He's from Virginia, and said they sent him here for some special mission. He wouldn't say more about that. He bragged about being a martial arts specialist and a sharpshooter, with special forces training." She stopped for breath. "I'm afraid I let him fool me."

"How so?" Weiss asked.

"We got involved, but after we went out for a while, he just ignored me. I wondered why they sent him to a headquarters finance section like ours, a big hotshot if you believed him."

"Anything else?"

"Yesterday, he got a long distance call, talked for an hour. Then he came out, all excited, and said we should report to the motor pool immediately. He issued us side arms." She laughed. "Our first lieutenant didn't want to ride in a truck. I didn't give a damn."

"What do you think the major is doing here?" Weiss asked.

"I don't know. He's got some big mission. Hush-hush. Very important. Oh yeah. The two sergeants with him?"

"I saw them," Weiss said.

"Watch them. They are his hatchet men. They would use their weapons."

My God, thought Weiss. It's worse than I thought. "See what's in that bag. You're probably hungry."

57

PROMOTION

Sheriff Weiss followed the convoy over the mountains the rest of the night. He decided to refer to his passenger not as the defector, but as the female second lieutenant who served as message courier from the military. She fastened her seat belt and passed out as soon as she finished Scotty's lunch.

Just after daybreak, they left the mountains on the outskirts of Red Valley. Weiss watched Chief Denning's female deputies meet the convoy ahead of him with their hats off. They talked casually to the people in the first few vehicles for a while, then led them toward the armory. Just now, the deputies may have saved a lot of lives, he thought. Sheriff Weiss dropped his passenger at police headquarters, told them that she was on "our" side, and could help.

Then he drove to the airport, parked, crawled into the pipe-rack bunk in the back of his vehicle, and passed out.

He came to with Chief Denning knocking on his window.

"Good morning, Chief." He used his electric razor as he talked.

"Morning Scotty. I think the plane is here." He pointed to a very unassuming military jet as compared to Air Force One. Two local rent-a-cars waited nearby. The jet stopped at some distance from any structures. The first person off the plane waved the rental cars out to him. He drove the first car to where the Sheriff and the Chief waited. A man jumped out, ran over to them, and shook their hands.

"You gentlemen are . . ." He came to a full stop and waited. Weiss introduced himself and the Chief, and showed his identification credentials. After carefully scrutinizing the papers and speaking into his radio, he suddenly became cordial.

"I'm an aide to the President. You probably realize this is an unusual visit. Would it be possible for my boss and two others to travel with you in your vehicle, while the Secret Service agents lead and follow in our rental cars?"

"Well, sure." My God, the Sheriff thought. The car's a mess.

"Can we go directly to see Dr. Jameson?" the aide asked.

"Yes. I can drive us directly to the shelter," Chief Denning said.

With that, the aide waved, and the other car pulled alongside. A man hopped out, and jumped into the back of the sheriff's vehicle.

He handed Chief Denning his Secret Service credentials, and said, "Gentlemen, we ask that you please follow any instructions we may give you at a later time." Then he moved to a jump-seat in the rear of the truck.

At that point, a man with a baseball cap pulled down, dark glasses, and a fleece-lined leather jacket with the collar up stepped out of the car and into the sheriff's vehicle, followed by another man and a woman.

A hand was thrust forward into the front seat. A familiar voice said, "Thank you for helping out, Sheriff Weiss."

Scotty couldn't help himself. He turned, shook the hand, looked into the familiar face of the President, and said, "Glad to sir. This is Police Chief Denning." Scotty noticed the President had on a bullet-proof vest. The President began the discussions immediately.

"May I introduce Fern Wilson, our legal observer, and Attorney General John MacLalen. I asked them to join us when we talked to you and Dr. Jameson.

"Gentlemen, some unusual events occurred recently in Washington. I know you are aware that Dr. Jameson, through the Burner Broadcast System, has presented news stories and documentaries that show corporate businessmen in an unfavorable light. The programs described how those businessmen mistreat us and the natural world." The President's voice was as persuasive in person as over national TV.

"Several corporation leaders told me that members of my own administration seriously disagree with my position. This is worrisome. It could mean that some of my own appointees, highly placed in the government, may have turned on me, and might try to harm me or Dr. Jameson. Have you learned anything about the people who are pursuing Dr. Jameson? I really need to find out who is behind this."

Sheriff Weiss then mentioned the new Hoopa shooting, and reviewed his impressions that the FBI agents seemed under extreme pressure to find Dr. Jameson and take her into custody. "If someone told me that the FBI Director himself had charged them to 'bring in Dr. Jameson or else,' it would make the agents' behavior more understandable. They are intensely disturbed and demanding. They are wild to find her." The Police Chief broke in.

"I agree. The agents seem desperate and dangerous."

"This confirms what I feared," the President said. "Whatever happens, Attorney General MacLalen is present to establish violations of law by the agents, certainly this new shooting. Do you think someone in authority might have charged the FBI agents with permanently silencing Dr. Jameson?"

"Yes sir," Weiss said. "I suspected it when both the FBI and the military showed up in Arcadia."

"It sounds very likely to me," Chief Denning said.

"Where are the FBI and the military groups now?" the President asked.

"They are resting at the armory in Red Valley," Chief Denning said. "I told them I had crews searching for Dr. Jameson, and would let them know the moment we learned of her location."

"So we have a little time, at least." The President sighed, and they drove in silence for a while.

Chief Denning told them about the tightly knit Red Valley community that harbored resentments against outsiders. Dr. Jameson received protection because of an extended family, almost a citizen-militia relationship among the rural residents of Red Valley and Humboldt County.

"At this moment, our progress is being followed by a civilian with binoculars and a cell phone." He laughed. "In fact, you can be sure that people are looking at us the whole time through telescopic sights on rifles." The Secret Service agent's jaw tightened.

"You will notice observers around the mine site. They would be there with or without my permission. Their concern is to prevent intruders from terrorizing a respected local resident. We are not vigilantes, Skinheads, or the Posse Comitatus." The Secret Service agent peered uneasily out the window.

Chief Denning led the cars to the mouth of the mine. All went inside on this sunny, dry California morning, except for three Secret Service agents who sought cover from rifle fire above and to the sides. They couldn't, of course, and felt exposed.

Two men emerged from the President's car with portable video camera packs. "Where do you want us?" they asked. The Police Chief waved to someone above them on the mountain. Soon a man with a rifle descended and escorted the cameramen back up the mountain in search of a vantage point.

While the President's men looked for defensive positions, Scotty Weiss received a telephone call. He hurried to the President with the news.

"Sir, I just got a call from some friends. I guess they are like the head of the local 'Neighborhood Watch?' On their way here they came upon a vehicle parked near your airplane. Two Mexican nationals inside couldn't explain their presence. My friends searched the vehicle and found a fully functional bazooka type rocket launcher. We fear they were waiting your return."

A Secret Service agent shook his head sadly. The President finally said, "What is worrying us, Sheriff, is how the hell they knew I was here. We have a leak in our organization. The people trying to kill me can find out where I am from moment to moment."

How the hell do you protect yourself when your enemy can keep you in his sights? Scotty asked himself, glad he wasn't in the President's shoes. But, of course, for the moment he was.

The bomb shelter was built in the mouth of an old mercury mine. The area around the mine's mouth was dry and sunbaked, with low, sparse vegetation. Remnants of roads, steel tracks, foundations for equipment and condenser tanks littered the mine site. It closed after WWII when the government dropped strategic materials subsidies. Many a miner picked out surface veins of mercury sulfide, heated the beautiful red mineral in a pot with a makeshift condenser, and died from mercury poisoning.

When the mine closed, the local citizens decided to use it as a bomb shelter in which to hide supplies in case of a nuclear bomb attack. The California state government offered classes on how to build shelters, but could not tell citizens who lived through the blast how to survive the long-term radioactive contamination of the world outside the shelter.

Red Valley citizens seldom found use for the mine shelter but kept it clean for display.

Since there was no view the mobile home's bright curtains covered the windows. A small table and chairs occupied the middle of the cheerfully lit living room-kitchen. When the President and his guests arrived, they crowded into the chairs facing inward. The Secret Service Agent searched the bedrooms and shower and placed his chair near the front door to control access. Colleen sat in a chair opposite the Secret Service Agent. Subdued, Evan served coffee to give himself a role with the group.

Until recently, Colleen ignored what politicians said, since they all accepted money to lie. She expected little from the President, but warmly greeted the Attorney General and the Law Clerk, Fern Wilson, a tall, dark-haired, serious woman. They all faced the middle of the trailer, as Evan served coffee. Colleen noticed that Evan dissembled flawlessly, a natural born actor. He excused himself to take coffee to the Secret Service men outside. The President began.

"Dr. Jameson, I had the pleasure, recently, of reviewing your background and accomplishments. You deserve congratulations on your Birnham Foundation award."

"Thank you, sir," Colleen said. "At the beginning, I never saw much potential in a hog breeding farm."

"My scientific advisors tell me we are in jeopardy from overpopulation, and the long-term starvation has begun," the President said. "The general public doesn't know much about that."

Colleen nodded. "True. And at the moment, the American

investment community is rubbing its hands at the coming chance for corporate farmers to make money feeding the growing number of starving people in the world. This is criminal. The effort to increase agricultural yields will cause faster erosion and loss of ground water." Colleen felt her anger rising. She very much wanted to tell the President exactly what the real problems were.

"When we run out of petroleum, our long term agricultural production will have been damaged, so even more people will starve. Also, there will be more people to starve due to the earlier efforts to keep hungry people alive. Our business community conceals this whole problem, which I find cynical beyond belief."

The President smiled. "I'm no longer surprised by self-serving deception. Let me tell you why we came here." He drank some coffee and continued. "You described in your newscasts the sort of legislation which could rescue us from corporate ownership of the media. The legislature must outlaw corporate contributions to election campaigns, and outlaw advertising, ultimately."

"I hope you agree," Colleen said. "Our critics tell us we are too harsh with the business community."

"No, no, no," the President said. "This brings me to the next problem. Corporations will fight with all their strength, since they risk losing control over almost all aspects of society. Their programming and newscasts tell us what to think and who to elect, their advertising tells us how to spend our money, and they hide the fact that they are depleting our resources for their profit."

The President looked at Colleen intently. "I don't want these people to silence you. Even more, I would like the public to see you as speaking for my administration. We wish you would accept an appointment, perhaps as Presidential Adviser now, or Secretary of Science at a later date? My biggest uncertainty is whether we could keep you safer in Washington, or where you are in California."

His offer startled Colleen. "I assume the public might not accept a woman, especially one my age, as scientific adviser."

"I think we can persuade them. In fact, you may have persuaded them already. The number of people who listen daily to your newscast on the BBS is among the highest of any program, ever. I hope you will consider joining us. We have a crisis on our hands." He stared at her while she wondered how one said 'no' to the President.

Evan entered the trailer in time to hear the President's last comment. As he passed Colleen's chair, he smoothly seized her hair, and jerked her to her feet. He slid into her chair and pulled her onto his lap as a shield. She sat in the line of fire between himself and the Secret Service agent.

"What are you doing?" Colleen asked, pulling away from Evan.

Evan showed them a pistol in his hand, and said, "Don't move." He pointed the gun at the Secret Service agent, who burst out, "Damn it, sir. Now I recognize him. I've done business with him before. He's FBI."

Evan put the barrel to Colleen's head and said, "Be careful. I'll shoot her."

The Secret Service man slid himself sideways until he partially shielded the President.

"Sir, you want me to jump him," the Secret Service agent asked.

"No, no," the President said. "Everyone relax. He mustn't shoot Dr. Jameson. She's the most important person in the room."

"That's smart." Evan smiled. "Maybe I won't have to hurt her."

58

While Evan held Colleen hostage in the shelter, Early Byrd appeared from a depression in the ground and frantically waved to the Secret Service agents outside the mine mouth. He ran toward them, shouting, "That guy who brought you coffee? He just called the FBI crew and told them that Dr. Jameson and the President are here. It looks like he has a cellular phone."

"God dammit," a Secret Service agent said. "We'd better get the Chief out of here."

"Wait, fellas. Think about it," Early pleaded. "You'll just find yourself in a high speed chase on unfamiliar highways. Please. Go on inside, bolt the doors, and get yourself one of those firing slots, see 'em? When the cars get here, we can cut 'em to pieces. There's 15 of us out here, and you three in there, shootin' through the doors. Those doors will stop .30/.30 slugs. We tried 'em out. I really think you should go on in. They'll be here in four or five minutes."

"I think he's right," one of the agents said. They ran inside the door and closed it. One of the agents approached the mobile home, shouting, "The FBI and the military vehicles are on their way here."

He opened the trailer door and found himself facing Colleen in Evan's lap, with Evan's gun pointed at him.

"Excuse me," he said.

The President calmly told him, "Go keep the others out."

"Right," the agent said, closing the door. He returned to the steel doors and reported that Evan had a gun on Colleen and the President.

"While we wait, what is it you want?" the President asked Evan.

"Well, as your man told you, I'm with the FBI," Evan said. "They assigned me to infiltrate the Population Institute. From what I've heard, it appears that the FBI Director would like to see both of you dead?" Evan asked, much amused.

Stunned at her underestimation of Evan, Colleen saw no hint of submissiveness. He had regained his sense of humor, and with just a few words, he taunted them, showed the President his control over them, and offered them a chance to bid for their lives.

"Pure paranoid speculation on our part," the President said, with a relaxed smile.

"Well you certainly convinced me," Evan said, enjoying his advantage. "Now tell me why I'd rather be a hero in my President's eyes, instead of an obedient employee in the eyes of my boss, the FBI Director?"

"Well, I can certainly do a lot more for you than he can," the President said.

"Good. Then let's wait and see how well the approaching storm troops do. If they lose, we can talk." Evan smiled and settled back. He slid Colleen's hips about on his lap, enjoying the feel of her thighs.

"Evan," Colleen said through gritted teeth, "you will regret this."

"Dr. Jameson, please don't be too disturbed," the Attorney General said. "If we live through the confrontation outside, we will need him alive when we go after the FBI Director. That is why Ms. Wilson and I are here."

"If you say so," Colleen said. She carefully shifted her weight on one side, trying to grind her hip bone painfully into Evan's lap.

"What is your name?" Attorney General MacLalen asked.

"Evan Rose," the infiltrator said. "Anyway, I'll keep her on my lap until the festivities are over." He kept the gun pointed at the Secret Service agent.

"Why don't you go see what is happening out there?" the President asked the Secret Service agent. "Tell our guys not to shoot our FBI infiltrator. Right, John?"

"Right," the Attorney General said, also speaking to the Secret Service agent. "Tell 'em to try not to kill anybody on the outside. It is very important that we have people to testify against the FBI Director in the coming weeks. You understand what I am saying?"

"Right, sir. I got it." He disappeared out the door.

"You're shivering," Evan said, pulling a comforter over Colleen's lap. Now he pressed one hand between her thighs and soon had an erection. He blew on the back of her neck and shifted her on his lap to better enjoy her.

"Evan, I don't care what these people say. I'm going to kill you."

Fern Wilson rolled her eyes.

Early Byrd talked to his sharpshooters as they lay prone, rifle slings taut, excited and ready to shoot out squirrels' eyes. A Secret Service agent jumped out of the steel doors, cupped his hands, and shouted, "My boss says please keep as many as you can alive to testify. OK?"

Early jumped up, waved at the agent and said, "OK. Got it."

The Secret Service man disappeared like a rabbit back into his hole. As he peered out of his firing slot, he said, "You have to admire the tactics of these people. I've never been able to defend the boss from behind a rifle-proof steel door before."

"I hope they don't bring a bazooka here."

A shout went up outside, and plumes of dust rose in the air as the trucks turned onto the dirt road leading to the mine.

Early talked to his friends on the radio. "The major and two sergeants are the most gung ho. They are driving the lead military truck. You can recognize the three of them from their blocked caps and bloused pants. I'm going to take my hailer and go up close where they can hear me. There's a gully right over there, OK?"

Early ran to his chosen spot and radioed the marksmen closest to the approaching caravan. As the four vehicles passed that group, the snipers shot out the tires of the last truck. The occupants jumped out and ran back down the road from where they came.

The sharpshooters stopped the second truck the same way. As the occupants collected back down the road, they were now non-combatants, out of the line of fire.

Early shot out the tires of the two FBI cars before they got within pistol range of the steel doors.

As he turned his attention to the Major's truck, it accelerated and wove its way toward the steel doors at the mine entrance. A hail of fire rained into the cab and the tires. Wounded, one of the sergeants fell out. The truck hit the steel mine doors at an angle, bending them in, but stopping the vehicle. The major and the other sergeant jumped out with rifles.

Early shouted anxiously above the engine noise, "They got grenades." A hail of bullets cut down the second sergeant before he reached the bent steel door. The Major made it through.

When he squinted inside for a target for his grenades, he found himself faced by two Secret Service agents.

"Drop it," they shouted. When the Major raised his rifle, the two Secret Service agents' shots knocked his legs backward and out from under him. The agents looked at each other in horror at how close the grenade had gotten to the President.

"Tourniquet him," one of the Secret Service agents told the other as he turned to the battle outside.

As the agents got out of the car, several put up their hands and threw their guns away. Five resolute men ran toward the mine entrance. "Drop 'em, don't kill 'em," Early shouted over his hailer.

Early and the Secret Service men inside shouted, "Halt. Stop, or we'll shoot." Two more agents fell with leg wounds.

When the three remaining FBI agents got close to the doors, they fired rounds toward the openings caused by the truck. Some of the rounds bounced back at them. As the agents reached the door, all the riflemen's shots converged on the same spot, and quickly cut their legs from under them.

The riflemen leaped up and ran to the FBI agents who had put their hands in the air. They searched the agents for weapons.

"OK, fellows, drop your pieces. Its all over. Drop them," Early shouted over the hailer.

Chief Denning and the Secret Service agents took charge of the FBI agents within reach. Early and his friends brought over the military personnel to help with emergency first aid. Chief Denning called in ambulances and police cars. A Secret Service officer talked determinedly to the wounded Major while the Attorney General listened. Fern Wilson asked questions of whoever was conscious.

Inside the trailer Evan gave up his gun to a Secret Service agent, sighed, reluctantly removed his hands from where he held Colleen, and let her stand. The agent held Evan's arms behind him.

Colleen turned to face Evan. "Why didn't you kill me when you had the chance?"

"There wasn't any hurry," Evan said, "and I hadn't had much sex for a while," Evan said. "You weren't that good, but you were available." He laughed. "And eager." He smiled contemptuously.

Colleen winced, took a step forward, and kicked Evan—shins, thighs, groin, stomach, anything she could reach. When she stopped, exhausted, the Secret Service agent held Evan up and asked, "You sure you're finished, Ma'am?" Reluctantly he let Evan slide to the floor.

Cheeks flaming, Colleen turned to the President. "Is that offer still open," she called.

"Absolutely," he said.

"Then I accept. Can I stay in Arcadia until you call me? I think I'm safe, here," Colleen said.

"I think you're right. You are safer here," the President said when they went outside and looked around at the rifle-bearing civilians who controlled the wounded, renegade FBI agents.

The President looked up on the hill and shouted to the cameramen, "Did you get all that?"

"Yes sir. Enough for a good movie." The cameraman happily shook his camera in the air.

The President talked to Sheriff Weiss, then to his own Secret Service agents.

Last, he went to find Early Byrd. The goddam liberal President stood Early at attention, kissed him on both cheeks, Foreign Legion style, and gave him a full military salute. Early managed to return the salute, thinking, I'm not even a Democrat.

The Attorney General supervised the collection of statements from everyone present. Fern Wilson stayed close, asked occasional questions, and took notes.

Someone finally noticed that Evan Thornton Rose had disappeared. Last they saw him, he lay gasping on the floor of the trailer where Colleen left him clutching his aching parts.

That night, the President asked Chief Denning, "Do you have a safe place where we can hole up for a few days?" The Chief installed them in the Civil Defense headquarters.

Until he found the FBI Director, the President felt safer here than in Washington D.C. He sent the video team to find the equipment they needed to take professional quality videotapes and rapidly disburse the images to broadcasters.

During the next few days, the Civil Defense headquarters whirled with activity. The military plane ferried people back and forth from Washington. Attorney General MacLalen sent most of the soldiers back to their base, sworn to secrecy and under threat of possible prosecution. He told the Base Commander to forget the incident, but if any superiors had questions, they should call his office in D.C. The Base Commander confirmed that he had routed a call to Major Deevers from a Brigadier General in the Pentagon.

The Brigadier General arrived shortly on the President's plane. He listened carefully to the deposition given by Major Deevers, telling how the General asked him to take soldiers to Arcadia to prevent Dr. Jameson from giving more broadcasts. The Brigadier General then admitted that the Chair of the Business Lobby was an ex-military friend who asked him to find someone like Major Deevers to stop Dr. Jameson. What did "stop Dr. Jameson" mean? He assumed it meant to incarcerate or kill her. These admissions were videotaped.

Next, the Attorney General determined that the FBI agents were recruited from the San Francisco Bay Area. He questioned the recruiting agent at length. This agent was highly placed, part of the Intelligence Division of the Washington, D.C. FBI office. He finally admitted he worked for the Internal Security-2 Section. This probably implicated an FBI Assistant Director immediately under Picketts.

The net began to close.

Attorney General MacLalen and Chief Denning made inquiries and decided that Evan Thornton Rose might be down in the mine behind the trailer where he was last seen. Denning arranged for a search crew to start looking for Evan.

Fern Wilson went back to her job as Supreme Court Clerk, with her head and notebooks filled with observations. If her boss Justice Suggs chose to ask her questions in the future, she could give him answers.

The President hoped he could complete all he needed to do before the national press corps found him. Attorney General MacLalen accumulated 3 tons of incriminating documents and video tapes. They asked Chief Denning to meet with them that evening in the Civil Defense headquarters, a steel-reinforced concrete tilt-up building built to resist earthquakes.

The Chief arrived at sunset. A Secret Service agent returned from the delicatessen with a bottle of bourbon and sandwiches. The President sloshed whiskey into four cups and offered a toast to Chief Denning.

"To you and your friends, Chief. You saved our lives."

"It's been a pleasure, sir. I didn't know there were any real people in Washington," Denning said.

"Television and newspapers bad-mouth government people," laughed Attorney General MacLalen. "The TV and newspaper owners are conservatives who want the public to think we are feeble-minded and can't make decisions."

"We're leaving tomorrow," the President told the Chief.

"Have your problems eased any?" Chief Denning asked.

"We have a lot more information on who is trying to get rid of us. But I think there will be more efforts to get me out of office," the President said. At that moment he glanced toward the door and saw a man pull out a gun and level it at him.

The President dropped to the floor as the man fired. The Secret Service agent nearest the gunman said, "Jesus," and grabbed the gunman's arm with both of his. He dropped his weight on the arm, and they both fell to the floor. Chief Denning reached them, and they disarmed and cuffed the man.

MacLalen reached down to the President. "Are you all right?"

"Yeah," the President said. "Sure glad I saw him, though." The Attorney General helped the President to his feet. They walked over to the downed gunman.

"Who are you?"

No answer.

"Who sent you here?"

No answer.

They examined his light brown coloring, his round face and eyes, full lips, long straight dark hair, short stubby fingers, medium height, and full build. His brown eyes became expressionless. He missed the shot and withdrew into himself. His pockets carried $3200 in assorted bills.

The President looked at his assailant. "My god, they're using undocumented aliens for everything. You have the right to remain silent, how does that go? You bastard." The President looked at Chief Denning. "Thumbscrews, Chief. I've got to know who is sending these people. Otherwise they'll just keep coming until one of them kills me."

"We'll work on him, sir."

The President left the Red Valley community which had served him so well and flew to a military base near Washington DC. There, he swore the Commanding General to silence.

59

Once again, the Chair of the Business Lobby called FBI Director Picketts as soon as he heard of the mine debacle in California.

"Picketts?"

"Yes sir. I'm sorry to tell you that . . ."

"Picketts, don't talk. Start wondering how you are going to live through this," the Chair told him.

Picketts heard further telephone dialing until a new voice sounded. The Chair of the Business Lobby returned his attention to Picketts.

"Picketts, there is a third party on the line, an old friend of mine, the Director of the World Bank. He is setting teams in motion who will arrange for the death of the President."

"Well, sir, that's not necessary," Picketts started.

"I told you not to talk," the Chair said. "The Director and I have a bet. He thinks his people will get the President before you do."

"A bet?" Picketts voice quavered.

"Yes. If his people beat yours, I've arranged for you to lose all your accounts, holdings, property, insurance policies, everything. At that point, your wife will be out on the street, your kids dumped out of college, and you will be turned over to the Attorney General with a full dossier on your illegal activities while Director of the FBI."

The Chair turned his comments to the World Bank Director. "This incompetent had both Jameson and the President confined in a cave in California. The 15 FBI agents and several military officers present managed to get themselves wounded, captured, and interrogated by a police chief, a sheriff, and a bunch of deer hunters.

"That's not very reassuring," the World Bank Director said. "Why not get rid of him? Isn't he a liability?"

"Probably," the Chair said. "I may have to do that. What are you going to do?"

"Well, I'll tell my teams they have to accelerate their schedules. It sounds as if the President may be able to take action against the FBI soon."

"Picketts, did you hear the Director's recommendations?"

"Yes, sir, I did," Picketts said.

"You'd better think up something new, quick," Picketts' old boss told him.

"Yes sir."

Something tore at Evan's eyelid, a sharp stabbing pain. He jerked his head to one side and his hand brushed away what bit him. Rat, he realized, and snapped himself upright and awake in the darkness. He remembered slowly that he had been captured by the President's Secret Service agents.

To escape, he had run full speed down into the darkened mine. He fell several times, and bruised his head repeatedly. Evan was in the dark and on the edge of panic. I must get oriented, he thought.

He felt about him and found a vertical timber which may have been what knocked him out. With the timber in one hand, he groped around and found the rock wall of the tunnel. He breathed deeply several times so that he could hold his breath and listen for any sounds. He only heard dripping water.

His fear became less when he felt oriented. His body decided he was standing in a tunnel stretching straight ahead. He gathered his courage, released the timber, and crept along the tunnel wall, one hand feeling ahead of him.

After some progress, he felt a rock give way just ahead of his forward foot. Funny, he thought. Then he heard the shower of falling rocks hit the shaft wall deep below him. Now his foot slipped into the shaft. He fell backward, frantically trying to avoid toppling forward into the shaft. The deep hole ahead of him frightened him so badly he couldn't move. He sat until he calmed himself. He felt forward and found the edge of the shaft into which he almost stepped. With his hand he followed its outline a few feet. Knowing its location reduced his fear a bit.

During the next listening period, he thought he noticed a very slight current of air on his cheek. He turned his head from side to side and located its direction. His spirits leaped as he realized that the air current led up the shaft and to his left, not back to the mine opening and the Sheriff's men.

He scrambled to his left keeping close track of the edge of the shaft. His enthusiasm at following the current of air lasted him for a half-hour. He noticed the mine floor became drier. He covered several hundred yards in his follow-the-wall crab stride.

Suddenly he stopped. Something was different but he couldn't tell what. In a few more yards, his hand encountered a plant stem. He stopped to listen. Very faintly, he heard a dog bark.

I'm out, he thought, and accelerated his pace, finding more plants. He could smell the outside. He anticipated seeing the sky and eagerly extended his arms. Suddenly his hand hit something hard. He felt all round it, and realized it was a heavy steel grating to keep people from entering the mine. On the edge of desperation, he felt entirely around the grate to see how it was attached. He traced a sturdy rim all around.

"No, please, I can't go back down there," he said. He took hold of the grate and shook it. Nothing moved.

He knelt and put both hands together on the grate's rim, and shook it back and forth. Nothing happened. He moved his hands around the rim a few inches and shook it again. He worked his hands around the rim before he noticed any movement.

He pulled dirt and stones away from the loose edge, and shook the rim back and forth. After several minutes of shaking and prying, the edge of the grate pulled inwards, and he opened a passageway big enough to squeeze through. He stood in the fresh air and looked up at the stars, then down to the horizon and some distant lights. He sat down and sobbed.

Gotta get competent, he thought, or they'll catch me again. He felt for his cellular phone. It was there and gave him a dial tone. He still had a chance. Evan crept down the mountain to the highway. He could hear the sheriff's crew at the mine entrance. Hiding in the underbrush, he dialed the cell phone. Pray there's enough current for one more call, he thought. Picketts' voice answered.

"This is Rose," Evan said.

"You're not supposed to call here."

"Special circumstances. Don't hang up."

"We thought you were caught," Picketts said.

"No, and I know where they are. I can finish it."

"Do it," Picketts said.

"I need some equipment and a new outfit. Have someone pick me up outside of Red Valley."

"Oh Christ, this time of night?" Picketts said.

"I'll wait an hour on the State Highway at the mine road. Have him drive by and signal," Evan said.

"Yeah, I know the drill." Picketts hung up.

When the car finally arrived, the driver blinked a flashlight. When he saw Evan, he slowed enough for Evan to jump in the back seat and bury himself under a blanket. The driver accelerated cautiously and was soon beyond the mine site.

The next day, Evan sat comfortably in a motel room in Sacramento with the FBI agent who picked him up. He had new luggage, clothes, a haircut with a tint, a hat, several thousand dollars, a box of rubber gloves, and some new weapons.

"The boss says if he catches you on cocaine again he'll shoot you himself."

"Not to worry. I don't use it anymore except with women," Evan lied. "Now get the fuck out of here. Leave the keys. I've got to call Picketts." He dialed and soon the familiar voice answered.

"Where are you? My man pick you up?"

"Yeah, I'm all set," Evan said. "Listen, how many people you got after Daddy at the moment?"

"We put out an open contract. By now, there might be a couple dozen looking for an opportunity."

"Ok, I'll join that bunch now. I expect to have access to Mama in a week or two, so I'll put her on the back burner."

"Yeah," Picketts laughed. "She's not worth as much either, is she."

"You want me to work on this or not?" Evan asked.

"Sure. I want to find out if you are as good as you claim you are." Picketts laughed and hung up.

Prick, Evan thought. OK, for that, I'll take out insurance. He called a lawyer he knew.

"Jason, just a quick request. I'm mailing you a letter on stationery from a Sacramento motel. Hold it unopened until I get back to you in two weeks? If something happens to me, then open it, OK?"

Jason asked several questions, to no avail. Finally he agreed to do it.

60

During the President's stay in California, his TV camera crew worked 18 hours a day, recording interrogations of wrongdoers that would shock TV viewers for years. The FBI agents involved in the University siege and the attack at the mine showed the sordid pressures pushing agents into illegal operations. Startling management failures and in-depth criticisms of the FBI administration emerged.

So many small old-boy networks existed within larger ones that a well-intentioned agent dared not use the formal FBI chain-of-command.

Colleen's BBS newscast programmers showed the attacks upon her, first at the University, then the shootout at the mine, which established her clearly as a David fighting the FBI Goliath. They showed many interviews with agents charged to "stop" her, agents set in motion by the business community.

After the shootout, a few dozen FBI agents moved into the J. Edgar Hoover building and refused to come out. Though not directly involved in the shootout, they did not want to answer questions from the Attorney General. Neither did their supervisors want them talking to the AG's office.

Colleen reported on the BBS that the President feared evidence would be destroyed in the Hoover building; the guilty avoided questioning and possible prosecution. Many letters to the BBS argued that the Hoover building should be opened up, "Like the can of worms it is," a writer from the West advised.

Congress still resisted passing the Free Elections and No Advertising bills. Legislators took little notice of Colleen's BBS newscasts until opinion polls began to reflect the new criticisms. Letters from the public turned more hostile than any congressional staffers could remember.

Six more states passed a version of Free Elections and No Advertising laws. The federal legislators from those states felt more pressure, and they became more vocal in the Washington debates. Finally, some California legislators offered a Free Elections and a No Advertising bill in the US Senate and House, and the formal battle began.

When the President left California, he hid in the east coast military base. He brought his Attorney General who explained the probable involvement of the FBI Director to the Base Commander. The Base Commander expressed little surprise. "I'd comment, sir," he told the President, "but on my promotion to General, I gave up swearing. What would you like me to do, sir?"

"I need to recover control of the FBI from the Director," the President said. "I asked some old friends in one of the military branches to form a task group. They will move when we have lawyer-proof evidence that the Director has defected."

The President also asked Judge Selwyn Ames, a well-respected, retired jurist from Virginia, to act as a Special Prosecutor who would review the evidence accumulating against the officers of the Business Lobby, the Pentagon General, and the FBI Director.

Special Prosecutor Ames concluded that they probably had enough evidence to prosecute several FBI underlings, Major Deevers, and the Pentagon General, but they needed more evidence against the FBI Director. "You sure as hell have grounds to fire him," Special Prosecutor Ames told the President.

They wanted to find Evan Rose. They hoped he would implicate Director Picketts.

The day after the Red Valley shootout at the mine, Early Byrd returned to his garage and tow trucks in Arcadia. He declined to chase Evan down the mine, claiming he was now a decorated military hero. He was even considering changing his political affiliation to Democrat. Early's tow truck operators cried scandal.

Chief Denning found three men who once worked the mine to try and find Evan. They buckled on explosion-proof helmet lamps and battery packs borrowed from the closest operating mine.

The mucker led the way, maybe because he was prepared for the unexpected. As the trio entered the mine entrance and walked past the mobile home, they reached the first timbering, then a large cavern.

"Might as well try to talk him out." After fruitless shouting, they spent another hour picking their way through treacherous irregular passages.

After they knocked loose a timber, and waited breathlessly for the roof to fall in, they returned to the surface. The mucker looked up at the sky and said, "By God, I promised my wife I'd never go underground again. I swear that's the last time."

The President faced a time constraint. Every day that FBI Director Picketts remained free, an assassin could kill the President and stop the investigation of FBI wrong-doing. The President had a sense of close personal danger, especially after the violence of the attacks at Red Valley. Each new military officer he met he eyed as a possible assassin. He didn't sleep well.

Waiting carried another disadvantage. Picketts might be busy removing incriminating evidence from the FBI files. Pentagon officers could be destroying information that connected them to illegal business activities.

On the other hand, if he moved against the FBI before he had convincing evidence, his credibility could be destroyed. The Republican Congress would gleefully try to impeach him.

The President decided he could wait no longer and brought the battle into the open. He and his military task group put the Hoover Building plan in motion on a Friday morning, at 10 a.m. An aide to the President asked to see the Superintendent of Police for Washington, D.C., in his office. The aide handed the Superintendent an envelope marked Confidential. It carried the Presidential Seal and White House return address.

The Superintendent opened the letter and read: "I ask that you meet my Assistant Mr. Denvoy in your office, at 1 a.m. tomorrow morning, Saturday, 15 hours from now, to receive important information. Would you be good enough to confirm this appointment with Presidential Assistant Denvoy at this time? I ask you to keep this inquiry confidential until we can explain the purpose. Thank you for your cooperation. The President."

The aide reached for a phone, and asked the Superintendent, "May I call Mr. Denvoy for you now, sir?" The Superintendent grunted, irritated. Damn politicians. They had nothing to do but try and look busy.

The aide dialed, muttered into the mouthpiece, and handed the phone to the Superintendent. "Presidential Assistant Denvoy, sir."

"Mr. Denvoy?" the Superintendent asked. "What is this all about?"

"I would like to confirm that you will meet me just after midnight tonight, one a.m., in your office."

"What if I say no?" the Superintendent asked, peevishly.

Denvoy laughed. "Oh, I'd have to chase down the President and ask him to please do my job for me. I'd rather not do that."

The Superintendent recovered himself. "That's not necessary. I'll be here."

"Thank you. And please tell no one about this until we meet," Denvoy said. The Superintendent hung up the phone. The aide smiled, thanked him, and left the Superintendent's office.

Presidential Assistant Denvoy appeared at the appointed time and woke the Superintendent as he napped on a cot. The Superintendent had spent many a night on the cot when things weren't going well in the District.

The Superintendent drew coffee from a nearby percolator. "So what's up?" he asked.

Assistant Denvoy opened out a map of the District around the government area that showed the access roads.

"Two hours from now, at three a.m., six mobile military units will appear at the locations shown and will travel along the routes marked until they reach the J. Edgar Hoover Building. They will surround, isolate, evacuate, and occupy the building, allowing access only to selected groups for at least several weeks."

"Do you anticipate any violence? Shooting?" the Superintendent asked.

"No. Of course, something unexpected could happen," Denvoy said.

"Who is running this show?" the Superintendent asked.

"The authorization is signed by the Secretary of Defense, the Attorney General, and the President," Denvoy said. "Therefore it outranks any other orders you might receive. If the Director of the FBI, Mr. Picketts, shows up, please take him into custody."

"That's it? What do you want of me, specifically?" the Superintendent asked.

"Keep the District running normally. You must honor the military units' blockades and perimeter lines. The military groups have been briefed by the Commanding Officer of their base. They have orders to fire any weapons necessary to go through road blocks. They will have artillery available to assure that they can carry out their mission."

"What?" the Superintendent asked. "Did you say artillery?" .

"I did say artillery. We will occupy the building even if we have to blow holes through everything that gets in the way," Denvoy said. "Superintendent, I want to thank you in advance for your wholehearted cooperation. Here are telephone numbers you may use to contact us. The troops will arrive sometime after two a.m. and will take the FBI Director into custody. Good bye, sir." Presidential Assistant Denvoy turned on his heel and left.

The Superintendent dialed his desk phone. "Honey, don't wait up. I'll be home late."

61

J. EDGAR HOOVER'S BUILDING

Just after 2:30 a.m., a small group of military vehicles clattered up to the night entrance of the J. Edgar Hoover Building, headquarters of the Federal Bureau of Investigation. This concrete mausoleum loomed ominously over Pennsylvania Avenue, on a line between the US Supreme Court Building and the White House, sitting somewhat closer to the political arm than the arm of justice, a critic once noted.

One truck reversed itself and backed a towed 105mm M102 cannon into position, its muzzle some 20 paces from, and pointing at the glass paneled entrance to the building. The crew jumped out, lowered the base plate to the sidewalk, and completed the loading drill. They crouched to one side, ready to fire. A squad of troops deployed on each side of the doors to control access. A group of Army officers, with a Major General in charge walked through the doors toward the security officer at the desk.

"Good evening, sir. What can I do for you," the young uniformed security man asked.

"Not much, son. Keep your seat. We're waiting for your boss, the Chief of Security for this building. He'll be here shortly. While we wait, would you be good enough to take a look outside at the artillery piece which is aimed at this lobby."

"Yessir, I was admiring it. Used to be the old M101 A-1 105 mm howitzer in Korea, upgraded to the Viet Nam M One Oh Deuce, sir."

"That's very observant, son," the General said.

"Thank you sir, but I should mention that if you fire that piece, the high explosive shell will take out us, this lobby, and the cannon crew as well, sir. Also, you can't stake the base plate into the concrete, and it will recoil clear across the street, sir. I'm sure you knew that, sir."

"Again, very observant, son. But only half right. The first round won't take out the crew, because it isn't an HE shell. We've loaded it with a Beehive round set to go off after it passes through the glass doors. Do you recall the specs on those shells, son?"

"Yessir," the security man said, visibly shrinking behind his desk. "It has a thousand or so fleshettes, those little darts to inflict wounds on ground troops. That's an anti-personnel device, isn't it, sir?"

"Good memory, son. So that gun will provide a one-shot answer to anyone here trying to resist our entry. We'll be coming in soon," the General said.

"Yessir. No problem, sir."

When the Chief of Security for the headquarters building arrived, the military commander required him to lead a team of soldiers to each room in the building and remove any people they found. Two Army officers were charged with the security of all rooms and the keys for those rooms. They detained all occupants and locked each room as they emptied the building.

By the next morning, an aide called the President. "When we entered the office of the Deputy Associate Director for Investigations, we found him without his clothing, investigating a woman also without clothing who claimed she worked in Intelligence Division."

"Not too intelligent to get caught like that. Bring 'em here," the President said. He assembled a "Truth Squad" to conduct interrogations at a nearby military base. Compromised, the employees might be more candid.

During the next few days, all FBI employees who worked in the building received a sheet of paper which urged them to tell the authorities of any special interest in Dr. Jameson, or the President, during the last couple of weeks. The announcement also said that FBI employees must come forward and report any illegal activities or programs of which they had knowledge. Such a report would do much to remove any stigma of guilt from the employee who reported these illegal activities. They flocked in.

The President told the press that he would discuss the takeover of the FBI building on Sunday. The moment he decided to make that move he asked Colleen Jameson, his new Science Advisor to join him at the military base. She and Jonah arrived from California, and Jonah went on to find a place to live near the new job. The President signed Colleen on as a staff member with a travel account, access to the limousine pool, security clearance, and other perquisites. Hampered by not having access to the White House, the President hoped to remedy that shortly. He and Colleen prepared the FBI takeover material for the nightly BBS newscast.

As she often did, Colleen opened the broadcast.

"Tonight, the BBS Newscast will show you a series of events with frightening implications, documenting illegal behavior of highly placed, appointed law officers. These events led the President to take control of the headquarters of the Federal Bureau of Investigation and all records in the J. Edgar Hoover Building in Washington. Here is Attorney General John MacLalen and Special Prosecutor Judge Selwyn Ames, to describe the sequence of events." Their faces appeared on a nearby video screen.

"Attorney General MacLalen, when did you first begin to worry about the FBI?" Colleen asked.

"The President called me a week ago Tuesday, after he attended a meeting with a group of corporate executives who call themselves the Business Lobby. Membership is only available to corporate chief executive officers. These are the most powerful men in America, all directors of US corporations. Later when I saw video clips from the meeting, I was struck by the angry tone of the comments these executives made to the President of the United States. See how you interpret the scene."

A couple in Plumtree, Kansas watched the news with their children. They saw the Business Lobby officers' eyes widen with emotion, the pounding fists, the flaring nostrils, and flying drops of spittle as the executives shouted at the President. The little girl asked, "They're not supposed to talk to the President like that, are they Daddy?" "I don't like it," her father said, "but who am I." The scene ended as the CEO shouted at the President, "You are out of touch with your own administration."

Attorney General MacLalen continued. "That last statement, 'out of touch with your own administration,' worried us. We feared that someone with whom the President works might have turned against him. After listening to all the data, we recommended that the President take measures to protect himself from an assassination attempt," the Attorney General said.

"Well, why didn't somebody make those people stop shouting at the President?" persisted the young girl in Plumtree. "I guess they are too big and strong," answered her father, searching for a way to explain "clout." "The police are afraid of them."

On the video screen, Colleen turned to Special Prosecutor Ames. "When did you get involved, Judge Ames?"

"The President appointed me to review the incriminating information as it accumulated. We tried to determine whether some members of the FBI sought to do harm to the President. Your viewers may have watched earlier scenes of the pursuit and attacks upon Dr. Jameson and the President. It became clear that we should investigate the FBI in such a manner that files in the headquarters building could be preserved. I suggested that the President seal off the J. Edgar Hoover building files, and bring in the FBI Director, Mr. Picketts for questioning. The President sealed off the building with help from the military," Special Prosecutor Ames said.

Attorney General MacLalen returned to the screen. "Now that we have sealed off the records in the J. Edgar Hoover Building, we need a system for analyzing them as soon as possible."

The President appeared next. "In that regard, I wish to announce the appointment of Dr. Colleen Jameson to the White House staff as

Scientific Adviser. Her leadership in establishing and overseeing these newscasts led us to seek her help in my administration. She graciously agreed. As her first responsibility, we decided she should oversee the examination of the FBI files. Dr. Jameson will tell you about those plans."

"Thank you, Mr. President," Colleen said.

The little girl viewing from Plumtree, Kansas said, "See, Daddy, that lady talks nice to the President." Daddy answered, "That's right, honey, and she got herself a job, too. Remember that."

Colleen introduced members of the new FBI Inquiry Commission, which would include members of the National Academy of Science, well known criminologists, political scientists, historians, and sociologists. The Attorney General's staff would provide immediate access to the files to gather evidence for the trials of the corporate executives, individual FBI agents, and soldiers who took part in the assassination attempt.

Colleen asked a representative of the American Association for the Advancement of Science to describe the FBI Inquiry Commission goals. She was a stern looking anthropologist, with long hair pulled back.

"We will, of course, bring in a broad array of scholars to examine the files to generate and test whatever hypotheses may emerge as we work through this archive. We plan to invite muckraking historians to answer certain disturbing questions which have accumulated over the years, such as, How much control did the FBI exercise over US political affairs? Were special files collected to intimidate prominent persons? Did the founding director's racial views and sexual preferences damage FBI operations?"

"All that?" Colleen asked, laughing.

"Indeed," the Chair said. "Society has waited long enough to see what improper powers emanated from that building. If the term 'muckraking' bothers the historians listening, then they shouldn't apply for access."

Colleen said, "I will argue that we should use our scholars to suggest and test trends, generalizations, hypotheses and theories. The results can help guide social planning and legislative decisions."

"Finally, we will submit the entire findings to systems analysts to decide what parts of the FBI are worth saving, and which should be eliminated."

"How do you think this effort will be received by the public?" the President asked Colleen.

"Well, I hope they think of it as zero-based budgeting for police bureaucracies. Let's see what should be retained. Think how much money can be saved cutting the FBI budget," Colleen said.

The President closed the newscast.

"Thank you, Dr. Jameson, for joining our team. Thank you for listening. You may hear criticisms of my investigation of the FBI, but do not judge me until you hear my side. We thank the BBS for this hour of uncensored news."

Fred Burner flicked his TV control to change channels. He spoke to his wife.

"My gawd, honey, who'd of guessed how Colleen's newscast would take off. It isn't even done well, technically. How much did you work with her on the programming?"

"We talked about general ideas, a few specifics. She's bright and takes suggestions quickly. I'm proud of her. Aren't you?" she asked.

"Well, sure, but she's making us money. We're likely to get program awards for several of her broadcasts. I'm running over with sponsor requests for big bucks. I don't even know if Colleen will agree. Ha!" He turned to his wife.

"Damn right I'm proud of her, and you too for helping."

PICKETTS

FBI Director Picketts decided he may have to try to kill the President himself. He wondered if he had any credibility left. He called the White House.

"Sir, it's Picketts, or someone with his voice."

The President grabbed the phone.

"Is that you?"

"Yessir, it's me. I wonder if you would let me come in without prejudice. Talk things over, give you my side of the case. Don't involve the Attorney General yet?"

"I'd like to see you, but we can't give guarantees."

"Would you meet me privately? Just the two of us? Talk over coffee?" Picketts asked.

"I've been dodging too many of your people's bullets to trust coming alone. I think you should turn yourself in to the Attorney General."

"Please reconsider. I've made mistakes, but . . ."

"Well, where would we meet?" the President asked.

"A downtown hotel?" Picketts asked.

They made an arrangement, and Secret Service agents waited, but Picketts never arrived.

The President concentrated on chasing down those who had set his would-be assassins in motion. University scholars dug with impunity through the files of the paralyzed FBI, finding astonishing materials which confirmed their darkest suspicions and made the general public decide they needed a new kind of federal watch dog. Each day brought out new peccadilloes, scandals, and double-dealing among the members of government and industry. Even during recent years, the FBI continued to collect information to make sure they could embarrass highly placed politicians whenever they needed votes, for example, on appropriations committees.

YOU'RE NEXT

Around the J. Edgar Hoover building, the troops solidified their position by erecting sandbag barriers at the entrances of the building and around the gun emplacements facing these entrances. Two of these howitzers on Pennsylvania Avenue stood immediately across from the Justice Department and the Internal Revenue Service building.

As April 15th neared, a bold midnight passerby managed to deposit two signs on the sandbags, visible to the occupants of the buildings across the street. The sign facing the IRS building said, "YOU'RE NEXT," while the one opposite the Justice Building said, "AND MAYBE YOU."

62

Von Foerster, and others, ask whether ". . . the time has come when man must take control over his fate in this matter and attempt to launch perhaps the most ambitious, most difficult and most grandiose enterprise in his entire history: the establishment of a global control mechanism, a population servo, which would keep the world's population at a desired level." (Umpleby)

Jonah and Colleen flew from Washington, DC into Arcadia and checked into an airport motel. She and Jonah would spend the next day at a Planning Committee meeting to approve Population Institute projects. Then Jonah would fly to San Francisco to visit Regis.

After dinner, as they tried to settle into the motel room, Jonah asked, "How do we keep up Institute momentum while you wander off politicking?"

"I wondered how you felt about my new job." Colleen said. "You think it's a waste of time?"

"No, no." Jonah said. "It's just that you're fun to work with, and I've had to work alone for a while. I'll tell you one thing, though."

"What?"

"It's a lot easier for me to get things done now that you are so well connected. So keep the job and let me complain. I just miss you," Jonah said.

Colleen hugged Jonah. "I prefer that. Thanks for telling me. Makes it easier for me to be away." She kissed him. "Why don't you operate the Population Institute from Washington? Work on anything you want?" Colleen asked.

"Well, that's really what I wanted to talk about. Do you think the birth rate will really decline fast enough with the virus out there in women's hands?"

"I don't know," Colleen said. "There are reports in Mexico City of super machos, guys selling their services to impregnate women who pay them. The church is trying to figure out if this is prostitution."

"Maybe Mary Henly was right," Jonah said.

"Hey. We agreed not to mention her any more."

Jonah persisted, "Yeah, but remember when you finally told her you had found a virus, she said, 'I hoped it was a hormone, something we could have turned over to the chemists to produce.'"

"So?" Colleen asked.

"Why shouldn't I work on a project to look for a chemical compound to back up the virus, for men and women?"

"Well sure, but where would you find a lab?" Colleen asked "Oh my god, I can do that out at work," she said, realizing her new status and sources of information.

"We'd only need a grant to cover our costs in a pharmacology lab near Washington," Jonah said.

"Sure. Places like Bethesda are crawling with research facilities. How about doing a long search of the literature to see where to start looking? Talk to the people on the Internet. You'll get more suggestions than you can use."

"Probably true," Jonah said.

"Let me ask around at work first," Colleen said. "I'll bet we can find a computer terminal and access to a big pharmacology library."

"If nothing else works, we'll foment a Manhattan Project on population control," Jonah said.

"Manhattan Project?" Colleen asked.

"In World War II, the Manhattan Project became an all out effort by the best minds in the country to solve a problem which threatened civilization. The problem then was fascists trying to develop nuclear weapons," Jonah said.

As they drifted off to sleep, Colleen said, "Jonah. I'm worried."

He roused himself. She seldom admitted to worrying. "What about," he asked.

"I like our projects. They seem manageable." She paused. "But we're not solving the problem unless we get the corporations stopped. They'll keep controlling the country and wasting resources until we're back living in caves."

"We need the legislation," Jonah said.

"That's right. I don't know if the broadcasts will do enough," Colleen worried.

"Can the President help?" Jonah asked.

"Who knows?" Colleen said. "Maybe we can influence him when we get back. We are still in the honeymoon period."

Jonah rubbed her back until she fell asleep.

JUST ROLL IT ON WITH YOUR TONGUE

Next morning, they met with Arcadia University members of the Institute Planning Committee. Always enthusiastic, the faculty asked questions and exclaimed over the changes since the last meetings.

"Anyone have a project that needs attention?" Colleen asked.

The University ecologist said he had been meeting with the educational technology section. They wanted to produce a correspondence course series to bring the business community up to speed on ecology and evolution.

"A crash course in the basics of the biological world and its ecological relationships?" Jonah asked.

"My very words," the ecologist said. "Actually all citizens need it, a general education course. They should be required to pass college level courses or exams in biology, ecology, and evolution, also taught through TV courses and self-study materials. Maybe we could ask that people complete the courses as a citizenship education requirement to get a driver's license? To hold public office? To open a checking account or handle anyone's money?"

"Love it," Colleen said. "Acting in an ecologically responsible manner becomes a basic social requirement.

Next, a public health science faculty member blew them away when she handed around a sheet which said:

"The Population Control Center of Arcadia State University will offer the Colleen Jameson Prize, a twice-yearly award of $25,000, to the person who most advances research toward one of the Institute's major objectives:

To describe, design, or develop a replicable behavioral intervention strategy or process which will bring about the interruption of reproduction in human females for several years."

Colleen was delighted. "Such a simple idea, directly from the Birnham Foundation."

The Public Health specialist said she found it an easy extrapolation from public health's emphasis on the culture of health behaviors. She suggested that presentations with merit could be published in a new journal, perhaps *The Behavioral Science of Reproduction.*

They approved the project and assigned a group to find the prize money.

The last suggestion for a project came from a very young and uncertain student, pushed by two friends into standing and making her proposal.

"I think you should offer an on-going contest with a $100 weekly prize for high school students across the country. The prize would be paid for the Best Idea of The Week which could reduce pregnancies and sexually transmitted diseases."

Jonah, always the champion of the students, loved it. "It's a sure winner. Only $5000 a year to focus all those inventive young minds on a problem they are thinking about anyway? I'm sure some of them will field test the procedures. We'll do it. How did you think of it?" he asked the student.

"Well, I don't know if I should talk about it. I thought of it when my girl friend told us how she gets her boy friend to use a condom."

The girl scuffed her feet as they waited attentively.

"She said she rolls it on with her lips and tongue. He doesn't seem to mind, that way."

Colleen took Jonah to the airport and put him on a plane to San Francisco.

GOODBYE TO THE FARM

As Colleen left the airport, she realized how much she needed this time off. About to dive into a new and intimidating job, she wanted a few days away, and she loved Arcadia, even in the fog. She wanted to say goodbye to her empty house and farm where she had hoped to be living by now.

Driving past the campus made her think of the time she'd spent with Evan there. Her thoughts flickered with occasional sparks from the encounter she found irresistible. Since the last meeting which showed his devious and dangerous personality, she wondered how she had been fooled so completely. Because he held a gun on the President he would spend the rest of his life in jail. Good riddance, she thought.

Every day, she looked at Jonah anew. He took on dimensions she had missed. She once saw him as a bit dull, but he was utterly reliable, and would listen to her concerns without jumping in immediately to try and solve her problem. Best of all, she enjoyed his company and conversation over all others.

On the way to her house, she saw a patrol car with red lights flashing behind her. She pulled over, wondering what she had done. Sheriff Scotty Weiss jumped out of the car and walked up to her window.

"Sorry lady, but I always stop the pretty ones so I can talk to 'em," he said, grinning. Suddenly sobered, he asked, "How did things go? I saw you last after that wild scene at the mine."

"Everything got straightened out, Scotty. I'm sure you saved my life, you know," Colleen said. "You probably heard that I finally got a job?"

"I did, and they're lucky to have you. Listen, I won't keep you. I just wanted to welcome you back." He took a step back, smiled, and saluted her. As she pulled away, Colleen felt a secure glow to be in a town where Scotty was the chief law man.

She slowly drove past the farm buildings, and remembered how little the teams accomplished of the intended permaculture operation. In her mind's eye, she saw the planned combination aquaculture ponds and hydroponic tanks, the movable poultry cages or "chicken tractors," the compost goats, the orchard tending St. Croix sheep, the ducks, geese, and a dog or two. That's what I miss, she thought. Where are my dogs and cats to say hello when I get home?

She needed some India Project notes and some cold-weather clothes she forgot. As she told Jonah, "a bunch of important and trivial stuff that I want with me."

Who knocked down the mail box? she wondered as she pulled in and got out of the car. Look at that. Somebody drove on the lawn and tore it up.

She unlocked the front door, and went in, with the sights from the house taking her back to the happy first few weeks she lived there. There was a big pile of mail, and a musty smell that permeated everything. She propped open windows and doors for ventilation.

Although it was hours before dark, she was weary. She went into the bathroom and washed her face. She would clean up tomorrow. She tilted down the toilet seat, and sat, sorting her thoughts and getting sleepier.

She went into the bedroom and dropped into an easy chair. She picked up the phone to call Regis and let him know she had arrived. She waited a moment and realized that the phone had no dial tone. Bother, she thought. I'll have to use my cell phone. She leaned her head against the chair to rest a moment, with a slight notion that something was out of order.

Colleen woke with a start, jerking forward off the chair's backrest. Night had fallen, and a cold wind blew across her face. She scrambled up to turn on a light and looked around the room. When she reached for the shotgun behind the closet door, it was gone. Where could Jonah have put it? Chilled, she rushed through the house, locking doors and windows. Cramped and aching from sleeping in the chair, she thought, Let's start over. Go in the bedroom, undress, and get into bed. She dreamed worried dreams.

63

SLEEP TIGHT

The first thing she noticed was a man's lips pressed against her cheek. She emerged from sleep, pushed down into the bed by someone sitting alongside her. "Who is it?" she gasped.

"Just me darling," a familiar voice said.

"Who is it?" Her voice cracked in fright and confusion.

He slipped under the covers, naked, and held her arm as he lay down. Colleen screamed, pulling and jerking away from him. With him this close, it burst upon her.

"Evan?" she cried.

"Yes, darlin'. It's me. I've missed you," he said. "I didn't mean those bad things I said. Don't kick me anymore. I'm in love with you."

Colleen pushed at him with her arms and knees, grunting with effort. She smelled alcohol on his breath.

"Aw, Colleen, don't fight me," he scolded her. Then he lost patience. He wrestled her over with her back to him, and bent her into a fetal position. He jammed her against the bed board, pried her buttocks apart, and maneuvered to enter her.

"Evan. Stop. I'll do it, but let me have something first. Some liquor. Don't you have something? Please, Evan. It won't hurt so much. You're awfully big, you know."

"You're damn right I am. They all say that," Evan chortled. Saliva drooled from the side of his mouth. "Well, why not?" He reached down to the floor and handed her a bottle of vodka.

Colleen wriggled her arms free and took a drink. She wanted something to quiet her panic.

She held the bottle to his lips and reached down for his penis.

"Ooh, that's nice, pussy and vodka," he said. "Hold on." He pinched some white powder from the bedside table and snorted it.

"More chaser," he demanded. Colleen forced a laugh, holding the bottle to his lips and pumping his penis.

"Ooh," he gurgled as the vodka went down, his face contorting with the pleasure of Colleen's hand on his penis. "I'm gonna' marry you, you know. You don't need that dip shit you got now."

She took another drink, and considered trying to brain him with the bottle. She held it to his lips again. Like a child he greedily sucked, still groaning with the pleasure of her hand.

"Don't stop," he said, reaching for an ampule also on the bed-side table. "Ooh," he moaned.

"What's that," Colleen asked.

"Popper," Evan said.

"Popper? You mean amyl nitrite?" Colleen asked. "For the heart?"

"Yeah," he said. "Heart, head, dick, whatever; the mind-blowin'est come you ever had."

"Show me how and I'll do it with you next time," Colleen urged.

"Right on," he groaned. As his orgasm started he clumsily broke the ampule under his nose. While he panted his way into the orgasm, Colleen scooped up several ampules and broke them a few seconds apart, cupping her hand over his nose to retain the fumes.

"Ah. Ah. Ahh." Evan twisted, his muscles tensed, he jerked over and over and twisted on top of her.

Colleen tried to move, but he growled and rolled his full weight onto her. Oh god, I can't stop him, she thought. He'll come right back at me any minute.

Her hands were pinned between her breasts, and she slowly pushed her thumbs up onto Evan's throat. Along each side of his Adam's apple she found the light flutter of his pulse. She gently pressed her thumbs into his neck, into the flutter.

Evan sighed and quivered for a time, and slowly grew quiet. Colleen increased the pressure on his neck for another two minutes. Evan was now having a stroke as Colleen's thumbs blocked the flow of blood through his cerebral arteries, cutting off blood and oxygen to the brain.

She started wailing softly. "Please, god. I didn't plan to do this. I didn't know what else to do." She wept quietly, mournfully, as she held her thumbs tight in position for several minutes.

Finally she pulled herself out from under Evan's senseless body and threw herself into the bathroom shower. When she emerged, she dressed and called Sheriff Scotty Weiss.

Colleen wept and pawed through Evan's clothes while she waited. She wondered what his name was this time. She found an envelope marked “To whom it may concern, open on my death.” Signed, Evan Rose. Colleen opened the envelope and skimmed the contents.

She quickly called the duty aide’s number for the White House. “This is Colleen Jameson,” she said. “Wake the President. He will want this immediately.”

When the President got on, Colleen told him that Evan was dead and had left a letter incriminating Director Picketts. “It says Picketts is hiding at the country retreat of a wealthy banker friend.” Colleen told them the name and address. “Maybe you should tap the phone for a while before you pick him up?”

“You’ve solved our two biggest problems,” the President said. “But you don’t sound very good.”

“I’m not. I’m going to stay with my brother Regis for a while.” She hung up, thinking, That’s not the way to treat the President.

To the Sheriff, Evan Thornton Rose looked as if he died of a drug overdose. The Sheriff was devastated that he hadn’t been able to save Colleen from what he assumed was a rape. She refused to make a complaint or a statement.

Sheriff Weiss found a taxi driver to take Colleen on the long trip to Regis’ apartment in San Francisco. Colleen paid the fare and slept through the night on the back seat.

The President decided that Colleen hung up the phone too abruptly. Something was wrong. He told an aide to find Jonah. The aide traced him to Regis’ phone number in San Francisco.

“Jonah,” the President said.

“Yes sir?” Jonah was sleepy and frightened.

“I just talked to Colleen. She’s on her way to San Francisco. ”

“I’m surprised. She planned to stay in Arcadia for a few days.”

“Well, she sounded unhappy. Jonah, I know you’ll take good care of her. I need both of you. Tell me if I can help, OK?”

“Thank you, sir. I will.”

Jonah’s apprehension grew as he waited. Some time after the sun rose, he heard a cab and met Colleen struggling up the stairway.

“Are you alright?” Jonah asked. Her face was lined and weary.

Colleen shrugged. “Sure.” She scarcely glanced at Jonah.

“The President called. He was worried about you.”

In the apartment Regis greeted her. “Want some coffee?”

“Yes.” She sat and stared at the cup he put in front of her. Then she looked up at Regis. “Jonah and I have to talk.”

“Right. I’ll be at the office.” Regis left quickly.

A chill went through Jonah. He sat across the narrow table from her and took her hand. "Can I help?"

She leaned her head on her arm. "Jonah, do you love me?" She spoke so softly he barely heard her.

"Yes, I do." Jonah waited.

"I have to tell you something." She winced and closed her eyes.

"You *want* to tell me something?" Jonah asked.

"No. I have to. You see, I . . . "

"Are you sick or injured?" Jonah asked, interrupting her.

"No, not exactly." She spoke very softly, shaking her head.

"Do you want to be here with me right this minute?"

She nodded yes, her eyes still averted.

"Will you take a few days off and spend them here with me?"

She nodded again. "I'd like that," she whispered.

"Do you enjoy talking to me as much as I like talking to you?"

"Yes. I love it." She looked a bit less tired.

"Are you willing to help me with the new lab work?"

"Of course! I want to." Her eyes rose to his.

"Shall we go and find a place to live near your new job?"

"Yes. I want a home to live in. With you." She nodded soberly.

"Am I the most attractive man you've ever known, and if not, will you pretend that I am?"

She hesitated a moment, then smiled. "Of course you are."

"Do you want to get married?"

Her head twitched. "No. Well, if you really wanted . . ."

"Want to have a baby?"

"Well . . . no," she gasped.

"Want to tell me why you're blushing?"

"No. I'm not." He saw her ear-lobes flush to match her cheeks.

"Is there still something you have to tell me but don't want to?"

She looked at Jonah, her eyes shining. She waited, conflicted, for several seconds, then made her decision. "No, nothing." She stood. "Come on. I'm exhausted." She led him into the bedroom.

They undressed and fell into bed. She lay looking at him. My god, he plays me like a cello. He strokes and I throb. She turned over and snuggled against him. "Could we sleep for a while?"

As they drifted off she asked, "Do you really love me?"

"I love you. Now get some rest."

"Well, I love you," she said.

64

JUST A CPA

Knowing Picketts' location, the Attorney General tapped the phones and recorded incriminating material for as long as it was useful. Then they surrounded the estate, and captured Picketts as well as other conspirators. When first interviewed, Picketts did not know that Evan Rose was dead. The early interviews were handled by Deputy Attorney General Searle.

"Mr. Picketts, Evan Rose told us you participated in a critique of an analysis paper involving Colleen Jameson," Searle said. "Is that correct?"

"I couldn't hope to remember all of the paper critiques I've been involved in," Picketts said.

"Is that a 'no'?" the Deputy AG asked.

"No," Picketts said. "It is an 'I don't remember.'"

The Deputy AG shuffled some papers. "We have a little trouble with that answer, Mr. Picketts. We asked Evan Rose whether you participated in the Colleen Jameson paper critique. Here's what he said." Searle read from a transcript of an interview with Evan:

"Picketts ran the meeting. Take me to the files in the FBI building. I'll show you the cabinet and folders with transcripts of the meeting."

Deputy Searle looked up from the transcript. "Do you really want to continue your claim that you don't recall such transcripts? That you don't recall the paper critique involving Dr. Colleen Jameson?"

Picketts put his head in his hands. "I don't recall."

"Dr. Jameson has been on television for several weeks and has become a prominent figure. You don't remember any discussions of her?"

Picketts muttered unintelligibly.

"I beg your pardon," Deputy Searle asked.

"I don't remember."

"I see," Deputy Searle said. "Mr. Picketts, I have the documents to which Evan Rose referred. They were in the files he described. Your signature is on the critique." Searle waited a moment. "Would you like to start over?"

Director Picketts put his head in his hands. "What the hell difference does it make? I didn't do all this." Picketts sounded disgusted.

Deputy Searle asked, "So who did all this if not you? A subordinate, without your knowledge?"

Attorney General MacLalen watched behind a one-way mirror.

"No, of course not," Picketts said.

"Who, then?" Deputy Searle asked.

"Mr. Searle, have you looked over my background closely?"

"No," Searle lied.

"Well," Picketts said, "let me review it for you. I started twenty seven years ago as a Certified Public Accountant with Wardren and Liu, Inc. They analyzed portfolios for investment risk, and they found out I was very good at it." Picketts sat up, squared his shoulders and sounded as if it felt good to talk for a change.

"Pretty soon they moved me, with Wardren's approval, over to Dextan and Fraley, Incorporated. They raised me slowly up the executive ladder until I became CEO. You know those companies?" He looked at the Deputy AG.

"I've heard of them. I'm told they are decision leaders and are well regarded," the Deputy said.

"Exactly so," Picketts said. "For a while I acted as a troubleshooter. The CEOs I got to know, especially those I served with on the board of the First Satellite Bank in New York, recommended me to dozens of their colleagues. I helped CEOs recover from unwise decisions."

The Deputy and MacLalen listened, engrossed, as the Mafia-like influence network unfolded.

"This made me one of the best connected people in the business," Picketts said. "I helped dozens of CEOs work through difficult periods. I came to know their most intimate blunders. Everybody trusted me and came to me for help. We were making a lot of money." Picketts lost his focus and stared away for a moment. "Lots of money." He shook his head and laughed.

"Then the voters made a wrong turn, and elected this goddam liberal President. We all disapproved and should have done something immediately."

"He's too liberal? But you knew that from the beginning," the Deputy AG said.

"Oh, we've worked around liberals before, no problem. They like money as much as conservatives. But this one has a genius for making the conservatives look like they made all the blunders. So we decided we had to get somebody to watch over him. Get some dirt on him, drugs, sex, whatever. Settle him down."

"That's when they moved you into the FBI?" Deputy Searle asked.

Suddenly, Attorney General MacLalen was in the room. "Just how did they accomplish that?" he asked.

"Well, we knew the Chairman of the Judiciary, a corporate lawyer. We worked up a plan to get me appointed and confirmed. We decided which Senator would suggest me to the President, and who would support me through the appointment hearings."

"So you headed the FBI for two years. Then what happened," the Attorney General asked.

"This Jameson woman started making wild allegations and wilder recommendations for legislation. We wanted her stopped," Picketts said.

"So you did ask Evan Rose to help you?"

"Of course. We had to do something," Picketts said.

"To quiet her? Kill her if necessary?"

"I hoped it wouldn't come to that." He dropped his head in dismay at having to admit this.

"But what about the attack on the President?" MacLalen asked.

"Well, the President put that Bat Masterson in the Treasury Department and set him on us." Picketts voice rose. "Masterson defaulted on the Treasury Bills of any company he didn't like. So Jameson would call the company an eco-polluter, or some such nonsense, and Masterson would default on their bonds." Picketts' voice got wilder.

"They kept confiscating our corporate assets." Picketts looked up in agony to MacLalen. "We couldn't allow that, for God's sake."

"Of course you couldn't," the Attorney General said. "Then what happened?"

"They pulled me out to talk strategy all the time," Picketts said. "They decided that the only way to get rid of our problems was to remove the President."

"This wasn't your idea?" MacLalen said, incredulously.

"Of course not. I'm a CPA, for God's sake," said the Director of the FBI. "It was those God damned friends of mine who worked me up the corporation chain of command."

"I don't believe you. You'll have to identify them," the Attorney General said.

The Deputy AG ran out of the room to bring in recorders and laptops. They talked long into the night. Late in the session, Picketts told how he met with several of his business colleagues who took him to a hush-hush meeting with the Director of the World Bank.

There, the World Bank Director told Picketts that the success of all the internal and overseas business activities of the U.S. depended upon this President being stopped. Picketts said,"They told me that if I could stop the President, then he, the Director of the World Bank, would assure me a life-long pension or a permanent job overseas."

During the next weeks, everything Picketts told them was checked and verified by investigators and the team of university scholars sent in to preserve the FBI records. They located and turned over volumes of incriminating material to the prosecutors.

The Attorney General and his assistants eventually worked four years, arresting and prosecuting their way up through the highest echelons of corporate America, by proving who had exercised undue influence on current and past FBI Directors, legislators, Presidents, and others. Some were convicted of attempting to overthrow the U.S. Government.

65

Free and untrammeled representation of the public is possible only when men and women in high office are not indebted to special interests for financial donations. Senate bill 3242, 84th Congress

In the days that followed Picketts' confession, the senate and house leaders refused to call the Free Elections and No Advertising bills to the congressional floor for a vote. The President reluctantly decided he had to try and tip the scales. He asked for a joint session of Congress. The law-makers could hardly refuse him since there had been attempts on his life, and daily he had exciting news of arrests as he and the Attorney General pursued the FBI and military to find who had commissioned his assassination.

The President arranged to address Congress in the afternoon. He refused to let the commercial news media into the proceedings, and only allowed BBS reporters and cameras to be present to broadcast the speech live.

In the wings behind the podium, the Cabinet waited for the last straggler to arrive. The President would not be upstaged. Their attire was selected for dramatic impact by a team of BBS TV makeup artists who now fussed over them all. Finally it was time.

The Cabinet members filed in to a row of chairs behind the podium, grumbling because two stars among them got special attention. The President was announced and he walked quickly to the podium in a quiet dark blue suit, nodding solemnly, accepting the applause.

Then he turned with a dramatic gesture, stepped back and waved in Bat Masterson, resplendent in a tuxedo with boutonniere, a large white carnation. The applause increased, along with boos, to which Bat responded with a boxers overhead two-handed grip.

Last, the President escorted Colleen to the podium, holding her hand as she appeared in a light blue suit and white blouse with lace collar. As she nodded to each side, the United Nations' Indian Women's service award around her neck glittered with subdued highlights, an inscribed hammered silver plate. The applauding legislators rose to their feet, women and supporters first, finally the entire group. The President graciously seated Colleen to his right with Masterson at his left.

Colleen watched the legislators settle into their chairs amidst jocularity and tension. The House and Senate Speakers rose and introduced the President, giving him permission to speak. A hush fell. The President stood behind a podium which Colleen hoped was bullet-proof. She examined his profile like millions of other voters, and decided he was more concerned and determined than ever before.

"Ladies and Gentlemen of the Congress and of the United States, thank you for this opportunity to talk with you."

"I must bring you news tonight, unpleasant news concerning a group of people living in the United States, but not really a part of our community. These are people who have violated our trust as fellow Americans. We thought they were part of us, but they manipulated us and cheated us for their personal benefit."

"There seems to be no practice too inhuman for them. They will do anything to control us for their profit and power." Colleen heard his intake of breath as he dove into battle.

"I refer, of course, to American big businessmen," the President said. Colleen heard gasps from the audience.

"I refer to the high-level corporation executives who continue to force upon us the disaster we call economic growth, without regard for any damages done."

"These people purchased the print and television media of this country to misinform us and keep our attention focused on growth, continuous, resource depleting, polluting growth." He spoke slowly, and Colleen remembered the films of Franklin Roosevelt with his inexorable delivery.

"We can think of these people collectively as a growth machine, a social cancer that grows on the body of society, eventually killing it, killing us all."

"Growth so profitable to them that they are willing to deliberately design products to break before they wear out, so they could sell us new ones, the disposable society."

"Growth so important they would lie to us and especially to our children through the brain-numbing medium of advertising that tries to convince by telling lies. How could you more quickly teach our young people contempt for the truth than to expose them to television advertising?"

"Growth that caused automotive executives to destroy the American mass transit system and force us to buy cars, buses, and trucks from them. Thereby they caused the combined disasters of deaths from accidents, pollutants which destroy health, and the misshaping of our society with roads and parking systems to accommodate this deadly dangerous transportation device, the automobile."

"Growth that paves over and washes away the soil that holds the plants that feed us. Growth that destroys the world's biomass and produces gases that heat up the cool air we breathe, that pollutes the surface and ground waters on which all life depends."

"Growth of the population of humans beyond any hope for their survival, so many that they destroy the planet's ecosystem that supports every living thing on earth." Colleen wondered how anyone could resist the logic, but she knew they did.

"The history of the United States shows how the continuing development of petroleum technology by the western world brought about all of these growth disasters. For over a century, industrial magnates forced development upon American law makers in the name of progress."

"We can afford no more of that, now." Colleen watched the President hold back his anger.

"Why have we tolerated it? we should ask. The answer lies in the question, Why do law makers accept campaign contributions?"

Colleen saw a ripple of resentment, growls, and muttering

sweep through the legislators. The President looked up, pretending surprise at these signs of disagreement.

"Ah, well, then the campaign contributions are not important to your legislative performance? Is that what you say?" the President asked.

"That's right," burst out an angry Senator from New York, dissuaded from leaping to his feet by his colleagues.

"Good." Colleen saw the President's knuckles turn white as he gripped the podium and thrust his body toward his listeners.

"Then you have nothing to lose by giving up those campaign contributions. You can vote in favor of the Free Elections law. It should save you much distress." Now, Colleen saw the President relax a bit.

"Did you know that the directors of corporate Political Action Committees recently described every last one of you in this room as being quite friendly and accessible to their approaches and arguments? They reported exactly how much money it takes to get access, to talk to each of you. You cost different amounts, you know. The range is from 350 dollars to 1500 dollars." The lawmakers laughed and groaned.

"Do you really like these paid PAC officers walking among you, acting warm and friendly, outgoing and nice, as if they owned the building, as if they stood on equal terms with you, even though *you* are the elected legislators? Do you like them putting their arms around your administrative assistants as if they were old friends of the family? As if they *owned* you?"

"Don't you feel guilty avoiding academic visitors because they are more knowledgeable, and demanding, and cranky than these lobbyists who are paid to be so nice and helpful?"

"Don't you resent dealing with these polished professional PAC officers who have their own staff to rewrite *your* legislative bills to suit their parent corporations? Do you like to send away a lone environmentalist who paid her own fare to talk to you because neither you nor she has the backup staff to fix a bad and unnecessarily complicated bill?

"How do you like approving bills that provide loopholes for special corporations, but are so complicated neither you nor the public can find those loopholes? The Japanese write short legislation. Why can't we write plain language bills?" Colleen tried to locate the listener who shouted, "That's right," in the audience. Others nodded.

"How do you like corporate executives who claim it is entirely appropriate for wealthy persons and organizations to use their money to gain extra access and influence from legislators?"

Two spoke. "No," and, "It's wrong."

"Do you like the fact that PAC Directors tell us there is no way to stop buying influence, that reform is impossible? That *you* don't care enough to change?" Colleen felt an angry tension fill the hall. The President raised his voice to its loudest volume yet.

"How do you feel about corporate claims that big government is the enemy, not big business? That *you* and I are the enemies of the voters, not these corporate PAC lobbyists?" The President paused, then thundered:

"Is that true about *you*?"

"No," many burst out. Now the President dropped his voice and waited a few seconds. Then he asked quietly:

"And how do you feel about the confessions we now have from corporate leaders who paid people to try to kill the President of the United States?" He gestured toward Colleen, "And this courageous woman seated before you?"

A low moan escaped the group, silent with pain. "Damn them," someone said.

"So I ask you to walk away from these liars. Free yourselves. Start now by adopting the Free Elections and No Advertising bills."

"I ask, Mr. Speakers, that you hold a vote now."

A chorus of agreement broke out.

The Speakers reluctantly shuffled forward, and polled the members. At the end, most chose or felt pressured to approve the bills. The President led the applause as the last vote was tallied.

The bills passed.

66

The President congratulated the Congress, and prepared to leave. "Walk with me," he said, and offered Colleen his arm. "See how the bastards treat you when you win one."

Colleen heard the commercial reporters ask if he would attend a press conference.

“Tune in to TV Free-America, the Burner Broadcast System newscast,” he told them.

Colleen was pushed aside while the President worked his way slowly out of the building, shaking hands with colleagues who wished to congratulate him. He hadn’t lost the touch that got him elected, and they felt proud of him.

As they stopped for photographs Colleen found herself looking at a man with dark glasses lifting a pistol toward the President. She screamed and ducked. Shots sounded and Secret Service agents converged on the wounded President.

The nation waited anxiously as the President emerged from surgery. Colleen and Jonah were among the first visitors because the President asked for them.

“How am I doing?” His voice was barely audible. He wanted to hear it from them.

“We talked to the surgeons, sir. You’re doing well,” Jonah said.

“You’re going to be fine, Mr. President.” Colleen tried to smile at him.

“Good,” he whispered hoarsely. “I want you to do something for me. It’s important.” He reached out and took Colleen’s hand.

“Of course.” They listened intently.

“Go back to work on the population problem. Find a solution. Will you do that for me?”

“Yes sir, we will,” Colleen said.

“We promise,” Jonah said.

“Good. So go, get to work.” Tired, the goddam liberal President dropped Colleen’s hand and closed his eyes.

Bartlett, Albert A., "Reflections on Sustainability, Population Growth, and the Environment," *Population and Environment: A Journal of Interdisciplinary Studies.*, Sept., 1994.

Brown, Lester, and others, editors, *State of the World--1984: A Worldwatch Institute Report on Progress Toward a Sustainable Society*, annual volumes from 1984 to present, plus occasional papers. NY: Worldwatch Institute, Norton & Co.

Catton, William R., *Overshoot.* Chicago: U of Illinois Press, 1982.

Caufield, Catherine, *Masters of Illusion: The World Bank and the Poverty of Nations.* NY: Holt, 432 pp, 1997.

Clawson, Dan, Alan Neustadtl, and Denise Scott, *Money Talks: Corporate PACS and Political Influence.* NY: Basic Books, Divisions of Harper Collins Publishers, 1992.

Cleveland, Cutler J., Robert Costanza, Charles A. S. Hall, and Robert Kaufmann, "Energy and the U.S. Economy: A Biophysical Perspective," *Science*, 31 Aug 1984, pp 890-897.

Clinard, M. B. and P. C. Yeager, *Corporate Crime.* NY: Free Press, 1983

Coontz, Stephanie, *The Way We Never Were: American Families and the Nostalgia Trap.* NY: Basic Books, 1992.

Daily, Gretchen C, Anne H. Erlich, and Paul R. Erlich, "Optimum Human Population Size," *Population and Environment: A Journal of Interdisciplinary Studies*, July, 1994.

Daily, Gretchen C, Editor, *Nature's Services: Societal Dependence on Natural ecosystems,* Covelo, CA: Island Press, 1997.

Domhoff, G. William, *Who Rules America Now?: A View for the 80s.* Simon and Schuster, 1986

Dye, Thomas R, *Who's Running America? The Bush Era*, Fifth Edition. Englewood Cliffs, NJ: Prentice Hall, 1990.

"EF! and the FBI: A Strained Relationship," *Earth First! Journal.* Dec 21, 1990.

Etzioni, Amitai, *Capital Corruption: The New Attack on American Democracy.* NY: Harcourt Brace Jovanovich, 1984.

Forsyth, Adrian, *A Natural History of Sex: The Ecology and Evolution of Mating Behavior.* VT: Chapters, Publ., 1993.

Gever, John, Robert Kaufmann, David Skole, and Charles Vorosmarty, *Beyond Oil: The Threat to Food and Fuel in the Coming Decades, Third Edition.* Niwot, CO, 80544: University Press of Colorado, 1991.

Hardin, Garrett, "Mythic Aspects of the Demographic Transition." *Population and Environment.* Fall, 1990.

Harris, Marvin, and Eric B. Ross, *Death, Sex, and Fertility: Population Regulation in Preindustrial and Developing Societies.* NY: Columbia University Press, 1987.

Helvarg, David, *The War Against the Greens: The "Wise Use" Movement, the New Right, and Anti-environmental Violence.* San Francisco: Sierra Club Books, 1994.

Humphry, Derek, *Final Exit: The Practicalities of Self-deliverance and Assisted Suicide for the Dying.* Eugene, Oregon: The Hemlock Society

Jackson, Wes, Wendell Berry, and Bruce Coleman, Editors, *Meeting the Expectations of the Land: Essays in Sustainable Agriculture and Stewardship.* SF: North Point Press, 1984.

Korten, David, *When Corporations Rule the World.* SF: Berrett-Koehler, 1996.

Lang, Tim, and Colin Hines, *The New Protectionism: Protecting the Future Against Free Trade.* NY: The New Press, 1993.

Molotch, Harvey, "The City as a Growth Machine," *American Journal of Sociology,* v. 82, no. 2, September, 1976.

Ophuls, Wm, *Ecology and the Politics of Scarcity.* SF: WH Freeman and Co, 1976.

Piven, Frances Fox and Richard A Cloward, *Poor People's Movements: Why They Succeed, How They Fail.* NY: Vintage Books, 1979.

Rich, Bruce, *Mortgaging the Earth: The World Bank, Environmental Impoverishment, and the Crisis of Development.* Boston: Beacon Press, 1994.

Ryan, Megan, and Christopher Flavin, "Facing China's Limits," in *State of the World 1995*, Lester Brown, editor. NY: Norton.

Savory, Allan, "Will We Be Able to Sustain Civilization?" *Population and Environment: A Journal of Interdisciplinary Studies*, November, 1994.

Seybold, Peter J, "The Ford Foundation and the Triumph of Behavioralism in American Political Science," *Philanthropy and Cultural Imperialism: The Foundations at Home and Abroad*, Edited by Robert F Arnove. Bloomington: Indiana University Press, 1980.

Tuchman, Gaye, "*Making News: A Study in the Construction of Reality.* New York: The Free Press, A Division of Macmillan Publishing Co., Inc., 1980.

Umpleby, Stuart, "The Scientific Revolution in Demography," *Population & Environment*, Spring, 1990.

Weaver, Paul H., *News and the Culture of Lying.* NY: The Free Press, A Division of Macmillan, Inc., 1994.

ORDER FORM

Lee Del Mar
Dr. Stuart Johnson
PO Box 2768
Aptos, CA 95001

Please send:

QUANTITY PRICE

______ copies of *The Aerosol Solution* @ $12.00 __________

Shipping is $3.00 for first book
and $1.00 for each additional book. SHIPPING __________

TOTAL __________

California residents add 8% sales tax=$0.96/book
Funds are enclosed as check or money order.

SHIP TO:

Name ______________________________________

Address ______________________________________

City/State/Zip ______________________________________

Phone Number______________________________________